PATTERNS IN MATHEMATICS

Investigating Patterns in Number Relationships

Written by Paul Swan

Published by Didax Educational Resources
www.didax.com

Order Number 2-165
ISBN 1-58324-154-X

C D E F G 10 09 08 07 06

395 Main Street
Rowley, MA 01969
www.didax.com

Foreword

We are surrounded by patterns. Patterns come in various shapes and forms, such as the patterns found in music, simple patterns on clothing and the more complicated number patterns found in mathematics. The number system we use is based on patterns. In fact, mathematics has been described as the "science" of patterns.

The activities in *Patterns in Mathematics* offer students the opportunity to participate in pattern work that introduces and develops the relationship between numbers. The activities predominantly focus on number relationships in tables and become progressively more challenging throughout the book. Activities range from completing addition tables, to number sequences, to "finding the rule."

Whether the students work independently or collaboratively on the tasks in *Patterns in Mathematics,* they will be developing their confidence to observe, describe, create and extend patterns.

Contents

Teacher Notes

Pattern is a fundamental element of mathematics.

"The mathematician seeks patterns in number, in space, in science, in computers and in imagination. Mathematical theories explain the relations among patterns ... Applications of mathematics use these patterns to 'explain' and predict natural phenomena ..." *(Steen, 1988 as cited in A National Statement on Mathematics for Australian Schools, 1991, p. 4)*

The word pattern, though, because of widespread use has taken on many different meanings and therefore the idea of pattern, even though fundamental to mathematics, has become blurred.

"A pattern is simply a repeated design or arrangement using shape, lines, colors, numbers, etc." *(DeKlerk, 1999, p. 97)*

People in general, and children in particular, seem to have a natural curiosity about pattern and a desire to make sense of the world. Consider how pattern assists in the development of language. Consider the role of relationships in a family. We can use this natural ability of children to help develop mathematical thinking. It seems that thefollowing four processes are fundamental to the learning of not just mathematics, but anything.

- Collect **DATA**
- **SORT** and **CLASSIFY** the data
- Find some **ORDER** among the data
- Discover a **PATTERN**

Whenever looking for patterns, it is important children pass through these stages. If a pattern that is forming is a little muddy, then we might need to collect some more data and sort and classify the data to find an order that does lead to the development of a pattern.

Patterns in Mathematics has been designed to formalize this process so students become more adept at spotting patterns. So often students are unable to complete problem-solving activities because they have either skipped one of the four steps outlined above or have failed to make sense out of the chaos. Whenever children "get stuck," encourage them to look at the data, perhaps collect more data, or sort the data in a different way in an attempt to find order.

This book only goes part way toward improving students' understanding of patterns. As teachers we need to help children appreciate the patterns in nature, architecture, music and time.

Give them strategies for sorting and classifying the data they collect. Teach the students how to use:

- *Tables*
- *Graphs*
- *Venn diagrams*
- *Carrol Diagrams*
- *Matrices*
- *Other sorting frameworks*

It is important in building up the students' appreciation of patterns that they are given the opportunity to:

- *create and build patterns (and describe the patterns they have created)*
- *copy patterns*
- *extend patterns*
- *find, describe and explain existing patterns*
- *find the missing element in a pattern*
- *show or describe the same pattern in a different way*

Teacher Notes

To help students discover patterns, teachers can encourage them to describe the patterns they see or hear by asking questions such as:

- *Can you continue my pattern? Show me.*
- *Can you make a different pattern?*
- *How is your pattern different from mine?*
- *Can you hear a pattern in the rhythm I am clapping? Join in when you can.*
- *Can you show (make, represent) the pattern using objects? Show me.*
- *Can you show the pattern a different way? Can you show it another way?*
- *What comes next? Why? How do you know?*
- *What is my rule? Explain. Describe.*

Eventually, students should be encouraged to describe a pattern using a rule. The rule may be described in may ways:

- *Concretely*
- *Verbally and later in written form*
- *Symbolically*

Students may lay down objects to describe a pattern in music, or explain in their own terms the relationship(s) they observe in a pattern. After being given the opportunity to gather their thoughts, the students could be encouraged to describe their patterns in words. Eventually, symbols may be used as a form of shorthand to streamline the written description of a pattern. Once the students have become comfortable describing patterns, the more formal use of symbols may be introduced. Students may move on to other symbolic ways of describing a pattern such as through tables, ordered pairs and graphs.

It is important for students to discover that patterns may be described in a variety of ways. Some students can "see" a pattern when the data are presented in one form but often can not when they are given in a different form. For example, relationships often become clearer when shown on a graph than when presented in a table.

Much of the focus of this book is on the development of number patterns which provide the foundation for algebra. It is important that children be given exposure to a variety of sequences (a pattern which follows a rule) in order for them to recognize patterns resulting from the data they have collected. Throughout *Patterns in Mathematics*, students are given exposure to number patterns in particular. These include:

- *Arithmetic sequences, where a constant amount is added to create the next term in the sequence. For example, 1, 4, 7, 10, 13, ..., where three is constantly added.*

- *Geometric sequences, where each term in the sequence is multiplied by a constant to produce the next term in the series. For example, 2, 6, 18, 54, ..., where each term is multiplied by three.*

- *Exponential sequences. These patterns grow rapidly and are often associated with patterns in nature such as the growth of cells, bacteria and so on.*

- *Recursive sequences, where each number in the sequence is a function of the preceding number, which in turn alters as the sequence progresses. The Fibonacci sequence is a good example of this.*

Icons have been included on every activity page to show the materials required for the lesson.

Teacher Notes

The ability to spot a pattern is an extremely important skill not only in mathematics but in the real world. Consider the day trader looking for patterns across hundreds of shares listed on the stock market or the medical researcher looking for the cure to a disease. Each is sifting through mountains of data, often with the aid of a computer. Each is trying to sort the data, trying to create order from chaos, in order to find a pattern. Even with the aid of the computer it is the human that needs to spot the pattern.

It is the ability to describe and generalize patterns that makes mathematics so powerful. With a few symbols, placed appropriately, very complex situations may be described clearly and concisely. $E = mc^2$ describes a complex relationship that many of us might not fully comprehend, but so elegantly shows the power of symbols to describe patterns and relationships. We need to encourage the children of today to notice patterns if they are to solve the problems of tomorrow.

Dedicated to Geoff White who helped to open up the world of patterns to me.

Paul Swan

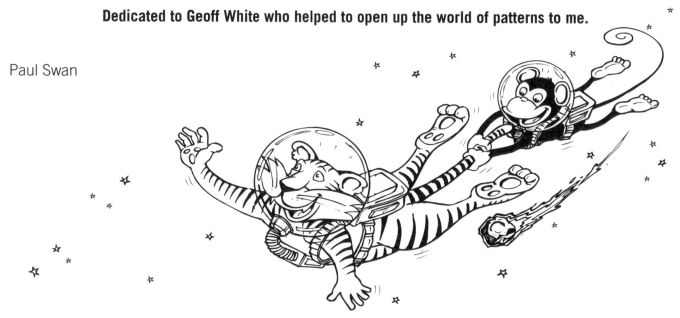

Using an Assessment Outline

How to use the outline is explained below.

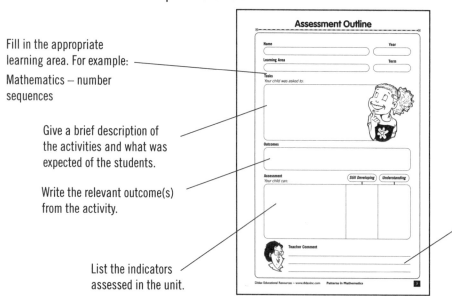

Fill in the appropriate learning area. For example:

Mathematics – number sequences

Give a brief description of the activities and what was expected of the students.

Write the relevant outcome(s) from the activity.

List the indicators assessed in the unit.

Use this space to comment on an individual student's performance which cannot be indicated in the formal assessment, such as work habits or particular needs or abilities.

Assessment Outline

Name

Year

Learning Area

Term

Tasks
Your child was asked to:

Outcomes

Assessment
Your child can:

Still Developing | *Understanding*

Teacher Comment

Addition Patterns

1

+	0	1	2	3	4	5	6	7	8	9
0										
1										
2										
3										
4										
5										
6										
7										
8										
9										

Complete this table of addition facts

= _____ 1st row

= _____ 2nd row

= _____ 3rd row

= _____ 4th row

2 (a) Add all of the numbers in the first row, second row, third row and fourth row.

(b) Write down any patterns that you notice. _____

_____

3 (a) Predict the results of adding the numbers in the

(i) 5th row _____

(ii) 6th row _____

(iii) 7th row _____

(iv) 8th row _____

I could use the answer from adding the numbers in the first row to help me work this out.

(b) Add the numbers in each row to check your predictions.

Check the answer if you predicted correctly.

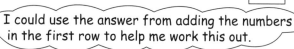

4 Predict the results for adding ...

(a) $10 + 11 + 12 + 13 + 14 + 15 + 16 + 17 + 18 + 19 =$ _____

(b) $12 + 13 + 14 + 15 + 16 + 17 + 18 + 19 + 20 + 21 =$ _____

(c) $10 + 20 + 30 + 40 + 50 + 60 + 70 + 80 + 90 =$ _____

More Addition Patterns

1 Complete this table of addition facts.

+	0	1	2	3	4	5	6	7	8	9
0					4	6	8	10	12	9
1										9
2										
3										
4										14
5										
6										
7										
8										
9										

9					54

9 =

9 + =

9 + + =

9 + + + =

4 + 6 + 8 + 10 + 12 + 14 =

This part is for the next activity

7th diagonal

8th diagonal

9th diagonal

10th diagonal

2

(a) Add numbers in each of the first six diagonals. Start in the top right corner.

(b) Describe any patterns you notice. _____

More Addition Patterns (continued)

You have only added the numbers along half the diagonals.

3 (a) Predict the totals for the remaining diagonals.

      (i) 7th diagonal _____

      (ii) 8th diagonal _____

      (iii) 9th diagonal _____

      (iv) 10th diagonal _____

(b) Add the numbers along each diagonal to check your predictions.

Check the answer if you predicted correctly.

4 (a) What do you think will happen if you continue to add the numbers along each of the remaining diagonals? (The next diagonal of numbers starts with ★.)

_____

_____

(b) Try adding this diagonal to see if your prediction is correct.

★   Total = _____   Were you correct? (Yes) (No)

5 (a) Investigate the sums of the numbers in the diagonals that run from right to left, starting from the top left corner. The first few have been done for you.

0	1 + 1 = 2	2 + 2 + 2 = 6	3 + 3 + 3 + 3 = 12

(b) Describe any patterns you notice.

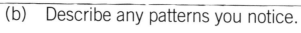

_____

_____

Patterns Within the Addition Table

1 Look at the following table of addition facts.

+	0	1	2	3	4	5	6	7	8	9
0	0	1	2	3	4	5	6	7	8	9
1	1	2	3	4	5	6	7	8	9	10
2	2	3	4	5	6	7	8	9	10	11
3	3	4	5	6	7	8	9	10	11	12
4	4	5	6	7	8	9	10	11	12	13
5	5	6	7	8	9	10	11	12	13	14
6	6	7	8	9	10	11	12	13	14	15
7	7	8	9	10	11	12	13	14	15	16
8	8	9	10	11	12	13	14	15	16	17
9	9	10	11	12	13	14	15	16	17	18

(a) Draw a rectangle around a block of four numbers.

e.g.

5	6
6	7

(b) Add all the numbers in the box.

e.g. $5 + 6 + 6 + 7 =$ _____

(c) Divide the total by 4.

e.g. _____ $\div 4 =$ _____

(d) Try the same thing with other blocks of four numbers.

(e) What do you notice? _____

_____

2 (a) Use the box below to look for a relationship that will allow you to work out the total of the four numbers in the box.

(b) Explain how your method will allow you to quickly work out the total of the four numbers mentally.

_____

3 (a) On the back of this sheet, investigate blocks of nine numbers.

(b) Describe any patterns that will help you to quickly work out the total of the

nine numbers. _____

_____

Challenge

Work out the total of the 100 numbers in the addition table.

More Patterns in the Addition Table

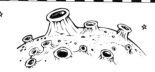

1 Look at the following addition table.

+	0	1	2	3	4	5	6	7	8	9
0	0	1	2	3	4	5	6	7	8	9
1	1	2	3	4	5	6	7	8	9	10
2	2	3	4	5	6	7	8	9	10	11
3	3	4	5	6	7	8	9	10	11	12
4	4	5	6	7	8	9	10	11	12	13
5	5	6	7	8	9	10	11	12	13	14
6	6	7	8	9	10	11	12	13	14	15
7	7	8	9	10	11	12	13	14	15	16
8	8	9	10	11	12	13	14	15	16	17
9	9	10	11	12	13	14	15	16	17	18

(a) Look at the cross shown on the table and write down the numbers in the top left corner and the bottom right corner and multiply them.

_____ X _____ = _____

(b) Multiply the numbers in the top right and bottom left corners of the cross.

_____ X _____ = _____

(c) Subtract the smaller answer from the larger.

_____ – _____ = _____

2 (a) Draw some crosses of your own on the addition table.

(b) Multiply the numbers in opposite corners of the cross and then subtract the smaller from the larger number.

_____ X _____ = _____   _____ X _____ = _____

_____ X _____ = _____   _____ X _____ = _____

_____ – _____ = _____   _____ – _____ = _____

(c) What did you notice? _____

_____

3 (a) Try drawing a larger cross on the addition table.
For example, a 4 x 4 cross would look like this:
Repeat the steps above for two different 4 x 4 crosses.

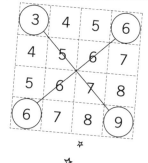

_____ X _____ = _____   _____ X _____ = _____

_____ X _____ = _____   _____ X _____ = _____

_____ – _____ = _____   _____ – _____ = _____

(b) When you multiply the numbers in opposite corners and subtract the smaller

number from the larger, what number is always produced? _____

Place Value Patterns–1

1   Try the following multiplications.

(a)   26 x 1 = _____

(b)   26 x 10 = _____

(c)   26 x 100 = _____

(d)   26 x 1,000 = _____

(e)   26 x 0.1 = _____

(f)   26 x 0.01 = _____

(g)   26 x 0.001 = _____

(h)   26 x 0.0001 = _____

A calculator will help.

2   (a) Describe the pattern you can see.

_____

_____

(b) In what ways are the answers the

same? _____

_____

(c) In what ways are the answers

different? _____

_____

3   (a) Describe what happens when you multiply by a number greater than one.

_____

_____

(b) Describe what happens when you multiply by a number less than one.

_____

_____

_____

4   Describe a method for doing calculations, like those above, in your head.

_____

_____

Then check your answers with a calculator.

5   Try these without a calculator first.

(a)   39 x 10   = _____

(b)   39 x 1,000   = _____

(c)   39 x 0.01   = _____

Place Value Patterns–2

1 Try the following divisions.

A calculator will help.

(a) 43 ÷ 1 = _____

(b) 43 ÷ 10 = _____

(c) 43 ÷ 100 = _____

(d) 43 ÷ 1,000 = _____

(e) 43 ÷ 10,000 = _____

(f) 43 ÷ 0.1 = _____

(g) 43 ÷ 0.01 = _____

(h) 43 ÷ 0.001 = _____

(i) 43 ÷ 0.0001 = _____

(j) 43 ÷ 0.00001 = _____

2 (a) Describe any patterns you notice. _____

_____

_____

(b) Describe a method for doing calculations, like those above, in your head.

_____

_____

_____

3 Try these without a calculator.

Check your answers with a calculator.

(a) 79 ÷ 10 = _____

(b) 79 ÷ 1,000 = _____

(c) 79 ÷ 10,000 = _____

(d) 79 ÷ 0.1 = _____

(e) 79 ÷ 0.01 = _____

(f) 79 ÷ 0.0001 = _____

Place Value Patterns-3

1   Try the following multiplications.

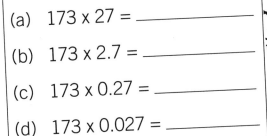

(a)   173 x 27 = _____

(b)   173 x 2.7 = _____

(c)   173 x 0.27 = _____

(d)   173 x 0.027 = _____

A calculator will help.

(e)   17.3 x 27 = _____

(f)   1.73 x 27 = _____

(g)   17.3 x 2.7 = _____

(h)   1.73 x 2.7 = _____

2   (a) Write about any patterns you notice. _____

_____

(b) In what ways are the answers the same? _____

_____

(c) In what ways are the answers different? _____

_____

(d) How does the position of the decimal point change the answer? _____

_____

3   237 x 16 = 3,792. Use this information to help work out the answers to:

(a) 23.7 x 16   = _____   (e) 237 x 160   = _____

(b) 2.37 x 16   = _____   (f) 2,370 x 16   = _____

(c) 23.7 x 1.6   = _____   (g) 2,370 x 1.6   = _____

(d) 237 x 0.16 = _____   (h) 0.237 x 160 = _____

4   Explain how patterns can help to work out answers to questions like those above.

_____

_____

Challenge

Write your own question similar to question 3. Begin with a multiplication fact, e.g. 52 x 13 = 676. Write six questions using these numbers. Change the value of the first or second number by moving the decimal point. Record the answers to the questions on a separate sheet of paper. Give your questions to a friend to work out.

    Patterns in Mathematics

Noticing Nines–1

There are many different patterns that may be found in the multiplication tables. Consider some from the nine times table.

1. (a) Complete the nine times table.

1 x 9 = _____		_____ 9 = _____ 9
2 x 9 = _____		_____ 1 + 8 = _____ 9
3 x 9 = _____		_____ = _____
4 x 9 = _____		_____ = _____
5 x 9 = _____		_____ = _____
6 x 9 = _____		_____ = _____
7 x 9 = _____		_____ = _____
8 x 9 = _____		_____ = _____
9 x 9 = _____		_____ = _____
10 x 9 = _____		_____ = _____

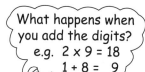

What happens when you add the digits? e.g. 2 x 9 = 18 1 + 8 = 9

What do you notice about the tens?

What do you notice about the ones?

What do you notice about the answers?

(b) Record your observations. _____

_____

_____

2. What happens if, instead of adding the digits in the answer, you subtract the smaller number from the larger? e.g. (4 x 9 = 36, 6 – 3 = 3.)
Describe the pattern that is formed.

_____

_____

_____

_____

Investigate here.

Noticing Nines–2

1 List the answers to the nine times table.

1 x 9 = _____9_____ (odd)

Write down whether the answer is odd or even.

2 x 9 = _____ ( )

3 x 9 = _____ ( )

4 x 9 = _____ ( )

5 x 9 = _____ ( )

6 x 9 = _____ ( )

7 x 9 = _____ ( )

8 x 9 = _____ ( )

9 x 9 = _____ ( )

10 x 9 = _____ ( )

1	8
2	7
3	6
4	5
5	4
6	3
7	2
8	1
9	0

2 When the digits in the answer are split, more patterns may be found.

(a) What happens when you add the digits along the diagonal? _____

_____

(b) Try adding the digits along the other diagonal; i.e.

8

2

Write about what you notice. _____

_____

3 What happens if you subtract the digits along each of the diagonals? (Remember to take the smaller number away from the bigger number.) _____

_____

4 Complete the following multiplications.

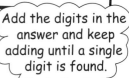
Add the digits in the answer and keep adding until a single digit is found.

(a) 736 x 9 = _____6,624_____ _____6 + 6 + 2 + 4_____ = _____18_____ = _____9_____

(b) 437 x 9 = _____ _____ _____ _____

(c) 615 x 9 = _____ _____ _____ _____

(d) 336 x 9 = _____ _____ _____ _____

(e) 167 x 99 = _____ _____ _____ _____

5 What do you notice when the digits are added? _____

_____

Patterns in Mathematics

The 99 Times Table

1. Complete the 99 times table.

2. You know about the patterns found in the 9 times table. Are there patterns in the 99 times table?

   (a) Discuss the answers with a partner.

   (b) What do you notice about the units, tens and hundreds digits?

   _____

   _____

   _____

   (c) What happens when you add the digits in the answer?

   _____

   _____

   _____

3. Try to write a rule to determine whether a number is divisible by nine without leaving a remainder. Share and discuss your ideas with a friend.

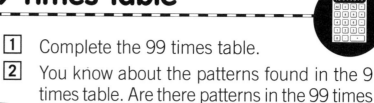

1	x	99 =	99
2	x	99 =	198
3	x	99 =	297
4	x	99 =	_____
5	x	99 =	_____
6	x	99 =	_____
7	x	99 =	_____
8	x	99 =	_____
9	x	99 =	_____
10	x	99 =	_____
11	x	99 =	_____
12	x	99 =	_____
13	x	99 =	_____
14	x	99 =	_____
15	x	99 =	_____
16	x	99 =	_____
17	x	99 =	_____
18	x	99 =	_____
19	x	99 =	_____
20	x	99 =	_____
21	x	99 =	_____

Eleven Times

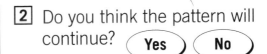

Do you know the pattern in the eleven times table?

11, 22, 33, 44, 55, 66, 77, 88, 99 . . . .

1 Notice what happens when the table is extended further.

(a) 11 x 11 = _____

(b) 12 x 11 = _____

(c) 13 x 11 = _____

(d) 14 x 11 = _____

Predict the following.

(e) 15 x 11 = _____

(f) 16 x 11 = _____

(g) 17 x 11 = _____

Check your answers.

4 Try these multiplications.

(a) 51 x 11 = _____

(b) 27 x 11 = _____

(c) 31 x 11 = _____

(d) 34 x 11 = _____

(e) 43 x 11 = _____

(f) 72 x 11 = _____

2 Do you think the pattern will continue? (Yes) (No)

3 Try some more to check.

(a) 18 x 11 = _____

(b) 19 x 11 = _____

(c) What happens when you reach 20 x 11?

_____

_____

_____

To multiply any two-digit number by eleven, take the two digits of the original number to form the first and last digits of the answer, e.g. 42 x 11 gives 4 _ 2. The middle digit is found by adding these two digits, e.g. 4 + 2 = 6 answer = 462.

5 What do you think you would need to do to multiply 68 by 11?

_____

_____

6 Now try these multiplications.

(a) 59 x 11 = _____

(b) 82 x 11 = _____

(c) 77 x 11 = _____

(d) 93 x 11 = _____

Puzzling Patterns–1

1 Complete the first three questions in each sequence using a calculator.
Use any patterns you discover to complete the rest of the questions without a calculator. Remember to check your answers.

(a) 99 x 12 = ☐

99 x 23 = ☐

99 x 34 = ☐

99 x 45 = _____

99 x 56 = _____

99 x 67 = _____

99 x 78 = _____

You can use a calculator here.

But work out the rest yourself!

(d) 4 x 2,178 = ☐

4 x 21,978 = ☐

4 x 219,978 = ☐

4 x 2,199,978 = _____

4 x 21,999,978 = _____

(b) 1 x 8 + 1 = ☐

12 x 8 + 2 = ☐

123 x 8 + 3 = ☐

1,234 x 8 + 4 = _____

12,345 x 8 + 5 = _____

_____ = 987,654

1,234,567 x 8 + 7 = _____

(e) 1 x 9 + 2 = ☐

12 x 9 + 3 = ☐

123 x 9 + 4 = ☐

1,234 x 9 + 5 = ☐

12,345 x 9 + 6 = _____

_____ = 1,111,111

1,234,567 x 9 + 8 = _____

12,345,678 x 9 + 9 = _____

(c) 9 x 1,089 = ☐

9 x 10,989 = ☐

9 x 109,989 = ☐

9 x 1,099,989 = _____

(f) 9 x 9 + 7 = ☐

98 x 9 + 6 = ☐

987 x 9 + 5 = ☐

9,876 x 9 + 4 = _____

98,765 x 9 + 3 = _____

_____ = 8,888,888

9,876,543 x 9 + 1 = _____

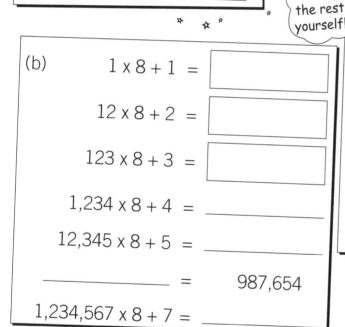

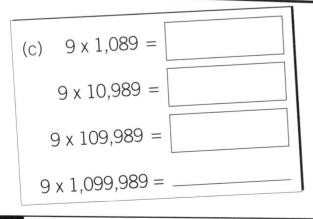

Puzzling Patterns–2

1. Complete the first three or four questions in each sequence.
Use any patterns you discover to complete the rest.

(a) 3 x 37 = ☐

6 x 37 = ☐

9 x 37 = ☐

12 x 37 = _____

15 x 37 = _____

18 x 37 = _____

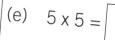

(d) 199 x 11 = ☐

299 x 11 = ☐

399 x 11 = ☐

499 x 11 = _____

599 x 11 = _____

899 x 11 = _____

Learning to use
the memory function
on your calculator
may help

(b) 7 x 15,873 = ☐

14 x 15,873 = ☐

21 x 15,873 = ☐

28 x 15,873 = _____

35 x 15,873 = _____

42 x 15,873 = _____

(e) 5 x 5 = ☐

15 x 15 = ☐

25 x 25 = ☐

35 x 35 = ☐

45 x 45 = _____

55 x 55 = _____

(c) 13 x 99 = ☐

24 x 99 = ☐

35 x 99 = ☐

46 x 99 = _____

57 x 99 = _____

79 x 99 = _____

(f) 265 x 9 = ☐

365 x 9 = ☐

465 x 9 = ☐

565 x 9 = ☐

665 x 9 = _____

965 x 9 = _____

Puzzling Patterns–3

1 Complete the first three or four questions in each sequence and use any patterns you discover to complete the rest of the questions.

(a) 3,367 x 33 = ☐

3,367 x 66 = ☐

3,367 x 99 = ☐

3,367 x 132 = _____

3,367 x 165 = _____

3,367 x 198 = _____

3,367 x 231 = _____

(b) 9,109 x 1 = ☐

9,109 x 2 = ☐

9,109 x 3 = ☐

9,109 x 4 = ☐

9,109 x 5 = _____

9,109 x 6 = _____

9,109 x 9 = _____

Now consider some division patterns.

If I learn how to use the memory key on my calculator I will save time.

(c) 1 ÷ 11 = ☐

2 ÷ 11 = ☐

3 ÷ 11 = ☐

4 ÷ 11 = ☐

5 ÷ 11 = _____

6 ÷ 11 = _____

7 ÷ 11 = _____

8 ÷ 11 = _____

(d) 1 ÷ 9 = ☐

2 ÷ 9 = ☐

3 ÷ 9 = ☐

4 ÷ 9 = _____

5 ÷ 9 = _____

6 ÷ 9 = _____

7 ÷ 9 = _____

8 ÷ 9 = _____

Challenge

Investigate what happens when you divide any two-digit number by 99. Use what you found to predict the result of dividing 35 by 99. Investigate what happens when you divide a three-digit number by 999. Now you can predict the result of dividing 123 by 999. Investigate what happens when you divide a four-digit number by 9999.

Staggering Additions

When one row of numbers is placed beneath another row of numbers and each separate column is added, some interesting patterns may be observed.

	1	2	3	4	5	6	7	8	9	10	
1	2	3	4	5	6	7	8	9	10	11	
			7	9	11	13					

\\/ \\/ \\/ \\/ \\/ \\/ \\/ \\/ \\/ \\/

___ ___ ___ 2 2 2 ___ ___ ___ ___

The difference between the number is found by subtracting the smaller number from the larger number.

1 Complete adding the rows in each column.

2 Find the difference between the numbers that are produced.

3 Now try the same with three rows of numbers.

		1	2	3	4	5	6	7	8	9	10	
	1	2	3	4	5	6	7	8	9	10	11	
1	2	3	4	5	6	7	8	9	10	11	12	
	3	6									33	

\\/ \\/ \\/ \\/ \\/ \\/ \\/ \\/ \\/ \\/ \\/

___ ___ ___ ___ ___ ___ ___ ___ ___ ___

4 What happens when you work with four rows?

			1	2	3	4	5	6	7	8	9	10	
		1	2	3	4	5	6	7	8	9	10	11	
	1	2	3	4	5	6	7	8	9	10	11	12	
1	2	3	4	5	6	7	8	9	10	11	12	13	
		6	10										

\\/ \\/ \\/ \\/ \\/ \\/ \\/ \\/ \\/ \\/ \\/

___ ___ ___ ___ ___ ___ ___ ___ ___ ___

5 Predict what will happen if you try the same idea with five rows. _____

_____

6 Test your prediction.

Challenge

Investigate what happens if the rows are staggered in different ways, for example:

1	2	3	4	5	6	7	8	9	10	
1	2	3	4	5	6	7	8	9	10	11
	1	2	3	4	5	6	7	8	9	10

Function Machines–1

When numbers pass through a function machine, they change according to the way the machine has been programmed. For example, the function machine below has been programmed to add five (+5). Note what happens when numbers pass through the machine.

The + 5 machine adds five

1. Insert the missing numbers for each function machine and explain what the machine is doing.

   (a) 5, 6, 7, 8, 9   + 10   **15, 16** _____   Adds 10

   (b) 100, 101, 102, 103   + 1   _____ _____

   (c) 10, 11, 12, 13, 14   – 1   _____ _____

   (d) 20, 30, 40, 50, 60   ÷ 2   _____ _____

   (e) 20, 30, 40, 50, 60   ÷ 10   _____ _____

   (f) 3, 6, 9, 12, 15   x 4   _____ _____

   (g) _____   – 5   10, 11, 12, 13, 14 _____

   (h) _____   x 2   10, 12, 14, 16, 18 _____

2. Explain how you worked out the answers to questions (g) and (h).

   _____

3. Try this function machine.

   _____   + 8   14, 15, 16, 17, 18 _____

Challenge   Write a "function machine" question and give it to a friend to answer. (Remember to write the answers on a separate sheet of paper.)

Function Machines–2

When numbers pass through a function machine they change according to the way the machine has been programmed.

1. Insert the missing numbers for each machine and give a brief explanation of what each machine does.

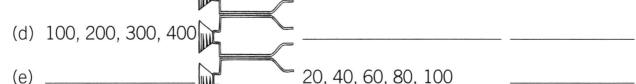

(a) 10, 12, 14, 16, 18 _____ _____

(b) 10, 12, 14, 16, 18 _____ _____

(c) 10, 20, 30, 40, 50 _____ _____

(d) 100, 200, 300, 400 _____ _____

(e) _____ 20, 40, 60, 80, 100 _____

2. Take a closer look at the numbers coming in and going out of the machines.

Write about any patterns you notice. _____

_____

3. Investigate what occurs when two function machines are placed together.

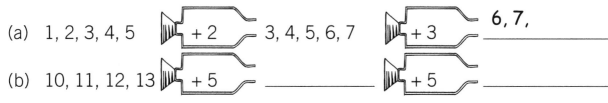

(a) 1, 2, 3, 4, 5 [+ 2] 3, 4, 5, 6, 7 [+ 3] *6, 7,* _____

(b) 10, 11, 12, 13 [+ 5] _____ [+ 5] _____

(c) What did you notice? _____

_____

4. Could you replace two machines with one?  Try these challenging questions to see if you are right.

(a) What happens when two subtraction machines are joined? Write questions of your own to find the answer.

(b) What happens when an addition and a subtraction machine are joined? On the back of this sheet, write questions of your own to find the answer.

Function Machines – Finding the Program

When a number is passed through a function machine it is changed according to the rule programmed into the machine. For example, the following function machine is programmed to multiply by 6.

This machine is multiplying by 6 !

[1] Try to work out what each of the following function machines does and write the program on the side of the machine. Explain in words what the function machine is doing.

(a) 1, 2, 3, 4, 5   4, 5, 6, 7, 8   Adds 3

(b) 1, 2, 3, 4, 5   11, 12, 13, 14, 15   _____

(c) 2, 4, 6, 8, 10   6, 12, 18, 24, 30   _____

(d) 10, 20, 30, 40   5, 10, 15, 20   _____

(e) 10, 11, 12, 13   5, 6, 7, 8, 9   _____

(f) 3, 4, 5, 6, 7   30, 40, 50, 60, 70   _____

(g) 3, 4, 5, 6, 7   0.3, 0.4, 0.5, 0.6, 0.7 _____

(h) 6, 7, 8, 9   7.5, 8.5, 9.5, 10.5   _____

(i) 10, 12, 14, 16   7.5, 9.5, 11.5, 13.5   _____

(j) 3, 6, 9, 12, 15   1, 2, 3, 4, 5   _____

Challenge

Write function machine questions with missing functions. Give them to a friend to solve. Remember to write the missing functions on a separate sheet of paper.

Swap the Numbers

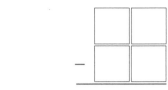
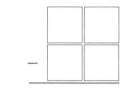

- Choose a two-digit number where the tens digit is greater than the units digit.
- Swap the two digits
- Subtract them
- Divide your answer by 9

42
24
$42 - 24 = 18$
$18 \div 9 = 2$

Write three examples of your own.

1 (a)

$\square\square - \square\square$ _____ $\div 9 = $ ___

$\square\square - \square\square$ _____ $\div 9 = $ ___

$\square\square - \square\square$ _____ $\div 9 = $ ___

(b) What pattern do you notice? _____

2 Further patterns can be found if you use numbers in sequences, e.g. 90, 91, 92, 93, 94, etc. There are nine two-digit numbers in the nineties where the first digit is larger than the second.

(a) List them all and calculate the difference.

$$90 - 9 \qquad 91 - 19 \qquad 92 - 29 \qquad 93 - 39 \qquad \square\square - \square\square$$

_____   _____   _____   _____   _____

What do you notice about the answers?

(b)

$\square\square - \square\square$ $\square\square - \square\square$ $\square\square - \square\square$ $\square\square - \square\square$

_____   _____   _____   _____

3 There are only eight two-digit numbers in the eighties where the first digit is larger than the second.

(a) On the back of this sheet, list them all and calculate the difference between the first and second digit.

(b) Write about what you notice. _____

_____

Challenge

Investigate what happens when you work with two-digit numbers in the seventies, sixties, fifties . . .

Patterns in Mathematics

Graphing Multiples–1

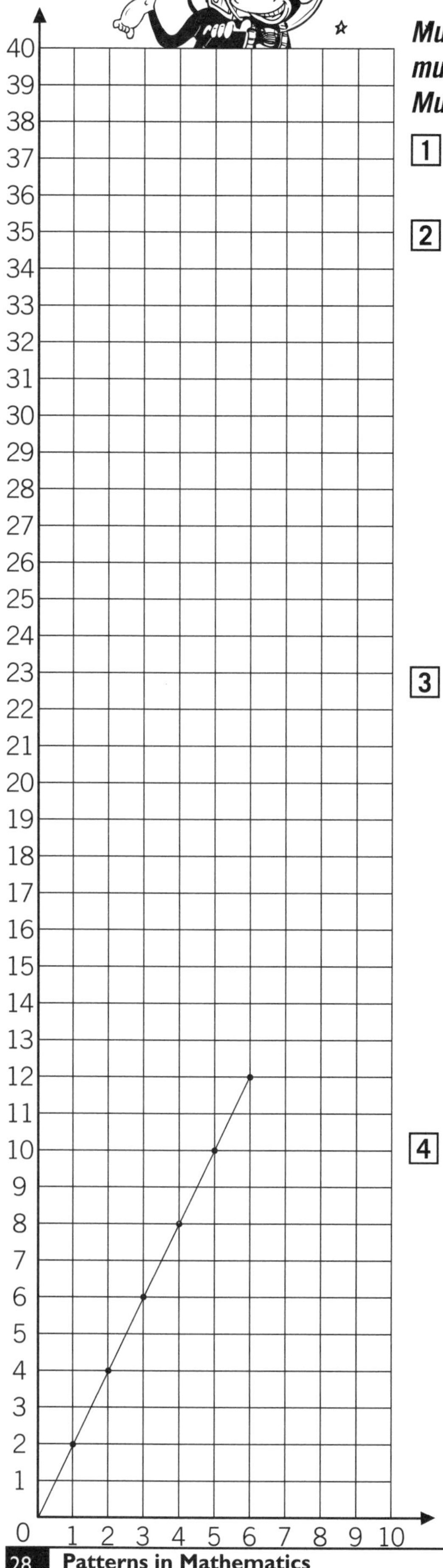

Multiples of two are 2, 4, 6, 8, 10, 12, ... The first multiple is 2, the second is 4, the third, 6 and so on. Multiples of two may be graphed as shown.

1. Mark in the 7th, 8th and 9th multiples of two on the graph and join the points.

2. (a) List the first ten multiples of three.

   _____

   (b) Graph the first ten multiples of three. Use a different color pencil. (Make sure it is sharp.)

   (c) Compare the graph of the multiples of two with the multiples of three graph. What do you notice?

   _____

   _____

3. (a) Describe what you think the graph of the multiples of four will look like.

   _____

   _____

   (b) List the first ten multiples of four.

   _____

   _____

   (c) Graph the multiples of four. Use a different color pencil.

4. Predict what you think would happen if you graphed the multiples of five.

   _____

   _____

Graphs are often a good way to find patterns and relationships.

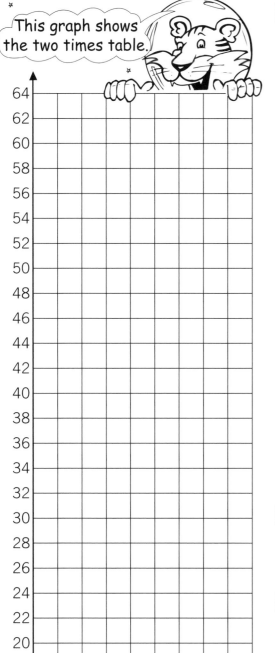

This graph shows the two times table.

1 (a) Complete this table of the multiples of 6.

1	6
2	12
3	18
4	
5	
6	
7	
8	
9	

(b) Graph these points onto the grid.

2 (a) Now complete this table showing the multiples of 7.

1	7
2	14
3	
4	
5	
6	
7	
8	
9	63

(b) Graph these points onto the grid. (Use a different color pencil.)

3 Which line has the steepest slope?

_____

4 (a) What would the graph of the one times table look like? _____

_____

_____

(b) Draw the graph to check your prediction.

(c) Were you correct?

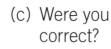

Yes

No

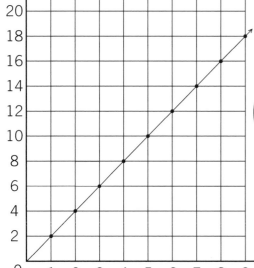

Marking Multiples–1

1 (a) Color all the multiples of five on the grid.

(b) Describe the pattern that is formed. _____

_____

2 (a) Use a different color and mark all the multiples of seven on the grid.

(b) Describe the pattern that is formed. _____

_____

_____

1	2	3	4	5	6
7	8	9	10	11	12
13	14	15	16	17	18
19	20	21	22	23	24
25	26	27	28	29	30
31	32	33	34	35	36
37	38	39	40	41	42
43	44	45	46	47	48
49	50	51	52	53	54
55	56	57	58	59	60

1	2	3	4	5	6	7	8
9	10	11	12	13	14	15	16
17	18	19	20	21	22	23	24
25	26	27	28	29	30	31	32
33	34	35	36	37	38	39	40
41	42	43	44	45	46	47	48
49	50	51	52	53	54	55	56
57	58	59	60	61	62	63	64
65	66	67	68	69	70	71	72
73	74	75	76	77	78	79	80

3 (a) Which multiples do you think will form a diagonal pattern on this 1 – 80 grid?

_____

(b) Mark the multiples on the grid.

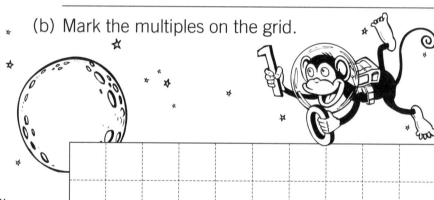

4 (a) Design a grid that will produce a diagonal pattern when the multiples of three and five are colored.

You don't need to use all of the grid.

(b) Test your grid by marking the multiples of three and five.

(c) How wide is your grid?

_____ columns wide.

Marking Multiples–2

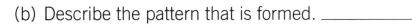

1 (a) Mark the multiples of three onto the 1 – 60 grid.

(b) Describe the pattern that is formed. _____

_____

2 (a) Which multiples do you think will produce the

same pattern on the 1 – 80 grid below? _____

_____

(b) Mark the multiples on the 1 – 80 grid.

1 – 60 grid

1	2	3	4	5	6
7	8	9	10	11	12
13	14	15	16	17	18
19	20	21	22	23	24
25	26	27	28	29	30
31	32	33	34	35	36
37	38	39	40	41	42
43	44	45	46	47	48
49	50	51	52	53	54
55	56	57	58	59	60

3 (a) List the multiples of six. _____

_____

(b) On a separate piece of paper, design a grid that will produce the same pattern when the multiples of six are colored.

(c) Test your grid by marking the multiples of six.

(d) How wide is your grid? _____ columns wide.

4 (a) Go back to your original 1 – 60 grid and mark in the multiples of eight using a different color.

(b) Which multiples would you need to mark on the 1 – 80 grid to produce a similar pattern?

_____

(c) List the multiples you would need to color on your own grid to produce a similar pattern.

_____

1 – 80 grid

1	2	3	4	5	6	7	8
9	10	11	12	13	14	15	16
17	18	19	20	21	22	23	24
25	26	27	28	29	30	31	32
33	34	35	36	37	38	39	40
41	42	43	44	45	46	47	48
49	50	51	52	53	54	55	56
57	58	59	60	61	62	63	64
65	66	67	68	69	70	71	72
73	74	75	76	77	78	79	80

Diagonal Dilemmas

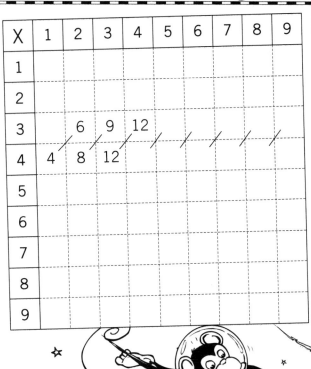

X	1	2	3	4	5	6	7	8	9
1									
2									
3		6	9	12					
4	4	8	12						
5									
6									
7									
8									
9									

1 (a) Complete this multiplication table.

(b) Shade the numbers in the third and fourth rows.

(c) Can you see the lines joining numbers in each row along a diagonal? Find the differences between the two numbers on each of the diagonals.

2, 1, _____

_____

_____

(d) What do you notice?

_____

_____

2 (a) Shade the numbers in the seventh and eighth rows. Mark the diagonals.

(b) Find the differences between each pair of numbers on the diagonals.

_____

_____

(c) What do you notice? _____

3 (a) Choose two more rows that are next to each other and mark in the diagonals. Find the differences between each pair of numbers.

_____

_____

(b) What do you notice? _____

_____

On a separate sheet of paper, investigate what happens if you change the direction of the diagonals.

Challenge

3	6	9	12	15	18	21
4	8	12	16	20	24	28

Multiplication Patterns

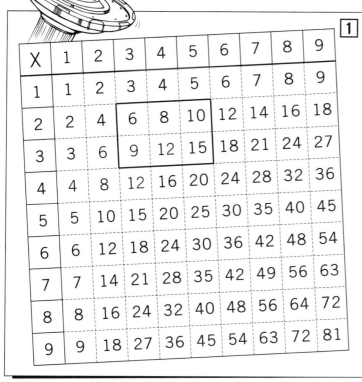

X	1	2	3	4	5	6	7	8	9
1	1	2	3	4	5	6	7	8	9
2	2	4	6	8	10	12	14	16	18
3	3	6	9	12	15	18	21	24	27
4	4	8	12	16	20	24	28	32	36
5	5	10	15	20	25	30	35	40	45
6	6	12	18	24	30	36	42	48	54
7	7	14	21	28	35	42	49	56	63
8	8	16	24	32	40	48	56	64	72
9	9	18	27	36	45	54	63	72	81

1 (a) Draw a 3 x 2 rectangle around six numbers in the table.
e.g.

6	8	10
9	12	15

(b) Multiply the two numbers in the opposite corners.

_____ X _____ = _____

_____ X _____ = _____

(c) What do you notice?

_____

_____

2 (a) Draw more 3 x 2 rectangles on the grid. Multiply the numbers in the opposite corners.

_____ X _____ = _____   _____ X _____ = _____

_____ X _____ = _____   _____ X _____ = _____

(b) What happens each time? _____

3 Draw different-sized rectangles on the grid and see what happens.

(a) 4 x 2 rectangles

_____ X _____ = _____   _____ X _____ = _____

_____ X _____ = _____   _____ X _____ = _____

(b) 3 x 3 rectangles

_____ X _____ = _____   _____ X _____ = _____

_____ X _____ = _____   _____ X _____ = _____

4  Predict what will happen if you have a 4 x 3 rectangle.

Now check and see.
Correct?   (Yes)
(No)

In the Middle

X	1	2	3	4	5	6	7	8	9
1	1	2	3	4	5	6	7	8	9
2	2	4	6	8	10	12	14	16	18
3	3	6	9	12	15	18	21	24	27
4	4	8	12	16	20	24	28	32	36
5	5	10	15	20	25	30	35	40	45
6	6	12	18	24	30	36	42	48	54
7	7	14	21	28	35	42	49	56	63
8	8	16	24	32	40	48	56	64	72
9	9	18	27	36	45	54	63	72	81

1 (a) Add the nine numbers in the first 3 x 3 block of numbers.

_____

_____ = ☐

(b) Divide the total by the number in the middle of the box. ☐ ÷ 6 = _____

(c) Repeat steps (a) and (b) for the other four 3 x 3 blocks.

_____ = ☐ ÷ 14 = _____

_____ = ☐ ÷ 25 = _____

_____ = ☐ ÷ 16 = _____

_____ = ☐ ÷ 64 = _____

(d) What do you notice? _____

2 *There is a faster way to find the total.*

(a) Mark in your own block of nine numbers. (3 x 3)

(b) Predict what you think the total of the nine

numbers will be. _____

(c) Explain how you made your prediction.

Have you found the short cut?

(d) Test your prediction.

_____ = _____ ÷ _____ = _____

The Reverse "L"

X	1	2	3	4	5	6	7	8	9
1	1	2	3	4	5	6	7	8	9
2	2	4	6	8	10	12	14	16	18
3	3	6	9	12	15	18	21	24	27
4	4	8	12	16	20	24	28	32	36
5	5	10	15	20	25	30	35	40	45
6	6	12	18	24	30	36	42	48	54
7	7	14	21	28	35	42	49	56	63
8	8	16	24	32	40	48	56	64	72
9	9	18	27	36	45	54	63	72	81

1 Add the numbers in each of the first four reverse "L" shapes.

```
1          2          3              4
        2  4          6              8
                   3  6  9          12
                             4  8  12 16
___        ___        ___        ___
```

2 Calculate the answers to each of the following multiplications.

$1 \times 1 \times 1 =$ _____ $2 \times 2 \times 2 =$ _____ $3 \times 3 \times 3 =$ _____ $4 \times 4 \times 4 =$ _____

> A number which is multiplied by itself is called a "square number."
> A number which is multiplied by itself and then multiplied by itself again is called a "cubic number."

3 What do you notice about the total for the first four reverse "L"s and the first four cubic numbers? _____

4 (a) Predict the sums for the next three "L" shapes.

_____   _____   _____

> You can color the "L"s different colors to help you.

(b) Explain how you made your prediction.

_____

(c) Check your predictions by finding totals for the 5th, 6th and 7th reverse "L"s.

5th _____   6th _____   7th _____

5 What do you notice about cubic numbers?

_____

_____

Robots–1

A toy robot may only be given two instructions.

- Turn right and

- Forward _____ steps.

The following instructions were given to a robot.

- Forward one step
- Turn right
- Forward two steps
- Turn right
- Forward four steps
- Turn right

These instructions are repeated until the robot gets back to the start.

A right turn is the same as a quarter turn or a 90° turn.

1. (a) Draw the robot's path.

   (b) How many times did you repeat the pattern? _____

start

2. (a) Follow these instructions to show the robot's path on the grid.

   - Forward four steps
   - Turn right
   - Forward one step
   - Turn right
   - Forward two steps
   - Turn right

   (b) Continue the pattern until the robot gets back to the start.

   (c) How many times did you repeat the pattern? _____

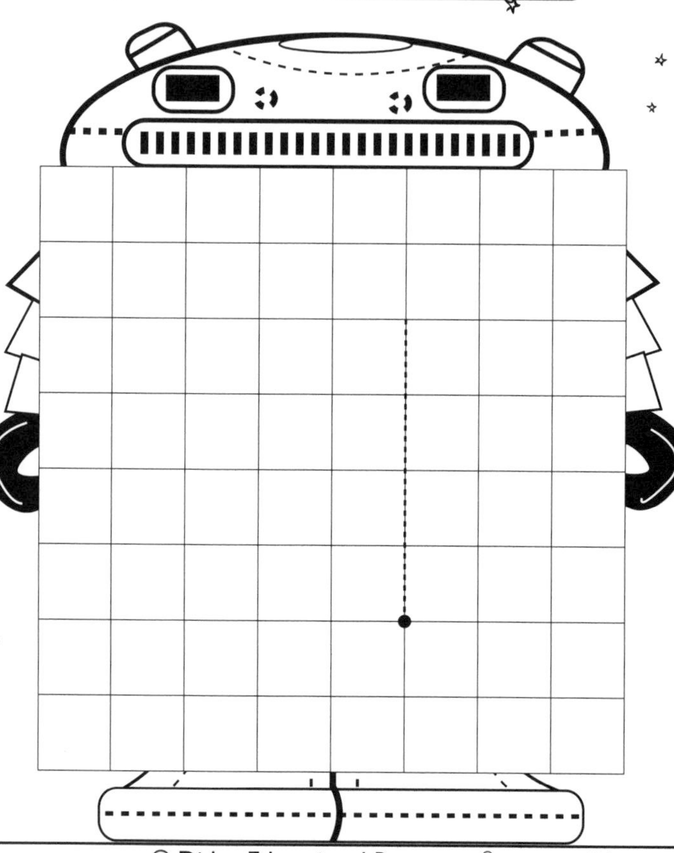

3. What do you notice about the patterns made by the robots? Write your ideas on the back of this sheet.

Robots–2

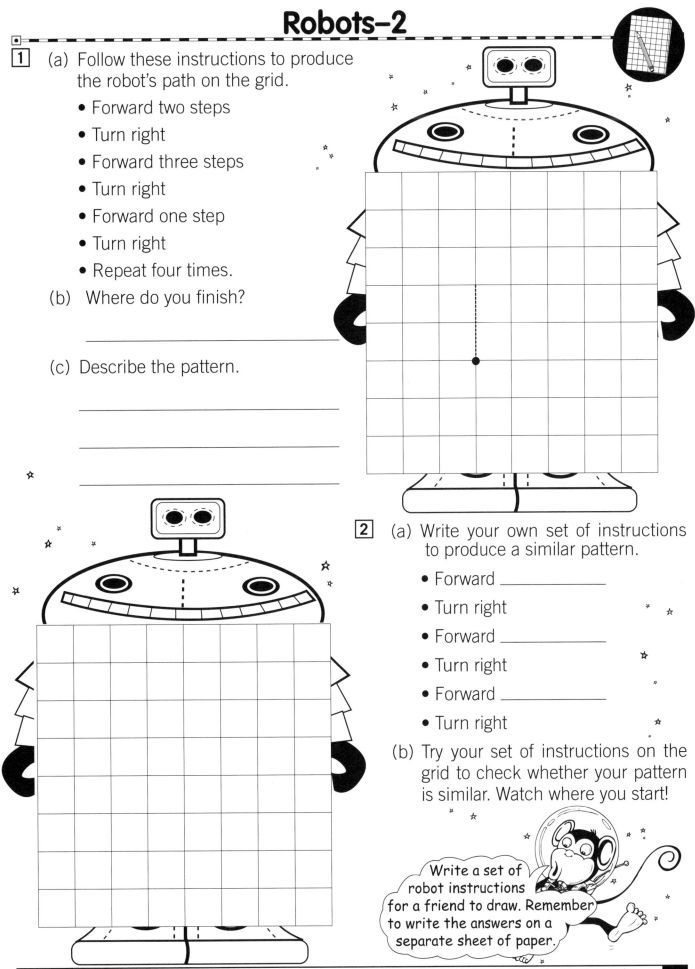

1 (a) Follow these instructions to produce the robot's path on the grid.

- Forward two steps
- Turn right
- Forward three steps
- Turn right
- Forward one step
- Turn right
- Repeat four times.

(b) Where do you finish?

_____

(c) Describe the pattern.

_____

_____

_____

2 (a) Write your own set of instructions to produce a similar pattern.

- Forward _____
- Turn right
- Forward _____
- Turn right
- Forward _____
- Turn right

(b) Try your set of instructions on the grid to check whether your pattern is similar. Watch where you start!

Write a set of robot instructions for a friend to draw. Remember to write the answers on a separate sheet of paper.

Even More Robots

You will need grid paper to experiment with different robot instructions.

1 Draw the paths the following robots would travel.

(a) Robot 1
- Forward 2 steps
- Turn right
- Forward 3 steps
- Turn right
- Forward 4 steps
- Turn right

(b) Robot 2
- Forward 3 steps
- Turn right
- Forward 4 steps
- Turn right
- Forward 2 steps
- Turn right

(c) Robot 3
- Forward 4 steps
- Turn right
- Forward 2 steps
- Turn right
- Forward 3 steps
- Turn right

(d) What do you notice about the paths traveled by the robots? _____

_____

(e) What do you notice about the instructions? _____

_____

> You may have noticed that the order of the instructions is the same in each case.

2 What happens if you change the order; for example, 3 steps, 2 steps and then 4 steps? Test what happens when the order is changed.

(a) Write your instructions and draw the robot's walk on the grid paper.

(b) What happens? _____

_____

3 What happens if the robots are instructed to walk 2 steps, 3 steps and then 6 steps?

(a) Write your instructions and draw the robot's walk on the grid paper.

(b) What happens? _____

_____

4 (a) Now try 2 steps, 1 step and 5 steps.

(b) What do you notice about the 2, 3, 6 pattern and the 2, 1, 5 pattern?

_____

_____

Circle Patterns

The following circle has been divided into 36 sections. The numbers closest to the circle go from 1 – 36 in a clockwise direction. The numbers on the outside go from 1 – 36 in a counterclockwise direction.

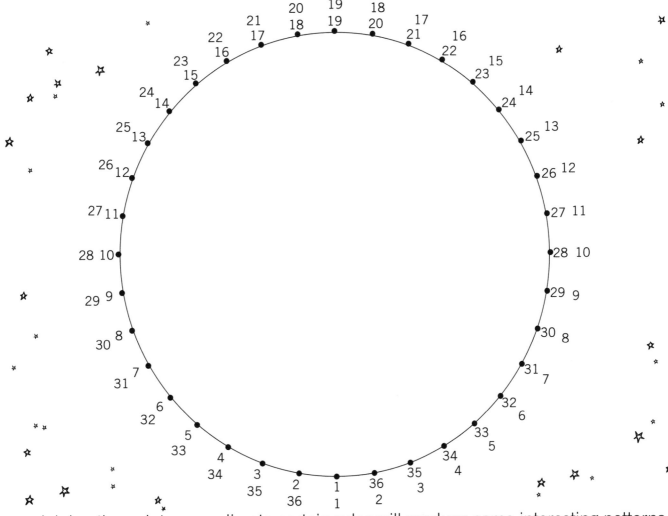

Joining the points according to certain rules will produce some interesting patterns. The first rule is n → 2n which means you will need to join each number to its double.   e.g. 1 → 2, 2 → 4, 3 → 6 and so on.

Join all the points to their double in a clockwise direction using a ruler and sharp pencil until you reach 18.

Now starting from 1 and working in an counterclockwise direction, join each point to its double until you reach 18.

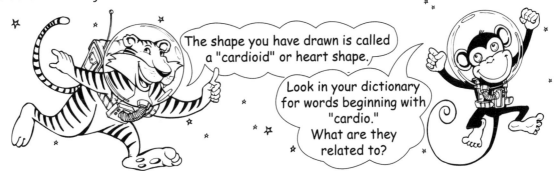

The shape you have drawn is called a "cardioid" or heart shape.

Look in your dictionary for words beginning with "cardio." What are they related to?

Digit Sums

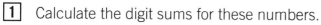

Every number has a digit sum. The digit sum of 7 is seven. The digit sum of 62 is eight (6 + 2). The digit sum of 728 is also eight (7 + 2 + 8 = 17, 1 + 7 = 8). The digit sum of a number is found by adding all the digits in the number until a single digit is left.

1 Calculate the digit sums for these numbers.

e.g. 38 → 3 + 8 → 11 → 1 + 1 = 2

(a) 41 → ☐ + ☐ = _____

(b) 382 → ☐ + ☐ + ☐ → ☐ → ☐ + ☐ = _____

(c) 4,886 → ☐ + ☐ + ☐ + ☐ → ☐ → ☐ + ☐ = _____

2 Many interesting patterns may be found by looking at the digit sums found in the tables.

(a) Calculate the digit sum for the six-times table.

			Digit Sum
1 x 6 =	6 →	6 →	6
2 x 6 =	12 →	1 + 2 →	3
3 x 6 = _____		→	_____
4 x 6 = _____		→	_____
5 x 6 = _____		→	_____
6 x 6 = _____		→	_____
7 x 6 = _____		→	_____
8 x 6 = _____		→	_____
9 x 6 = _____		→	_____
10 x 6 = _____		→	_____

	Digit Sum
11 x 6 = _____ → _____
12 x 6 = _____ → _____
13 x 6 = _____ → _____
14 x 6 = _____ → _____
15 x 6 = _____ → _____
16 x 6 = _____ → _____
17 x 6 = _____ → _____
18 x 6 = _____ → _____
19 x 6 = _____ → _____
20 x 6 = _____ → _____

(b) Write about any patterns you notice. _____

_____

(c) Predict the digit sums for the:

(i) 21st multiple of six. _____   (ii) 25th multiple of six. _____

(iii) 31st multiple of six. _____

3 Find the digit sums for the multiples of 2, 3, 5, 7, 8 and 9 and look for multiples with similar digit sum patterns.

_____

_____

Adding Consecutive Numbers

What happens when you add consecutive odd numbers, starting at 1?

1 (a) Try some.

Consecutive numbers means following one after another.

$1 + 3 =$ _____

$1 + 3 + 5 =$ _____

$1 + 3 + 5 + 7 =$ _____

$1 + 3 + 5 + 7 + 9 =$ _____

$1 + 3 + 5 + 7 + 9 + 11 =$ _____

When the information is placed in a table, a pattern emerges.

(b) Complete the table.

No. of Odd Numbers	Consecutive Odd Numbers	Total
1	1	1
2	1 + 3	4
3	1 + 3 + 5	
4	1 + 3 + 5 + 7	
5	1 + 3 + 5 + 7 + 9	
6		
7		
8		
9		
10		

Do you recognize these numbers?

2 (a) Predict the total for adding the first twenty consecutive odd numbers.

_____

(b) Explain how you made your prediction. _____

Growing Squares

1 (a) Continue this pattern. Note: only count small squares.

1　　　**2**　　　**3**　　　**4**　　　**5**　　　**6**

(b) Enter your data into the table below.

Length of side	1	2	3	4	5	6	7	10
Number of squares	1	4	9					

(c) Explain how you worked out the number of squares for a side length of 10. _____

_____

2 (a) What is the side length of a pattern made up of 225 small squares? _____

(b) Explain how you worked out your answer. _____

_____

(c) If you had 150 small squares, what would be the side length of the largest square you could make? Work with a friend to find the answer. Use the back of the sheet.

Answer _____

Visual Patterns

The following arrangements were made on pegboards.

8 pegs on the outside

1 hole in the middle

12 pegs

4 holes

16 pegs

9 holes

1 (a) Draw the next pattern in the sequence.

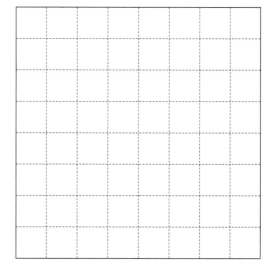

(b) With a partner, complete the table and predict the next two numbers.

Pegs on the outside	8	12	16	20	24	28
Holes in the middle	1	4	9			

(c) Check your table by drawing the next two patterns in the sequence.

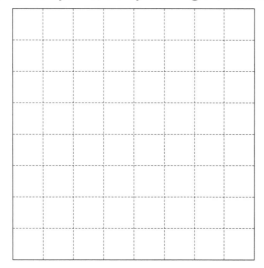

(d) Write about any patterns you notice. _____

_____

2 (a) If there were forty pegs on the outside, how many holes would you have? _____

(b) Explain how you found your answer. _____

_____

Multiplication Squares

X	1	2	3	4	5	6	7	8	9
1	1	2	3	4	5	6	7	8	9
2	2	4	6	8	10	12	14	16	18
3	3	6	9	12	15	18	21	24	27
4	4	8	12	16	20	24	28	32	36
5	5	10	15	20	25	30	35	40	45
6	6	12	18	24	30	36	42	48	54
7	7	14	21	28	35	42	49	56	63
8	8	16	24	32	40	48	56	64	72
9	9	18	27	36	45	54	63	72	81

1 (a) Starting at the top left corner of the chart, draw successively larger squares.

(b) Find the total of all the numbers in each of the first four squares.

(c) What do you notice? _____

_____

_____

2 (a) Predict what you think will happen for the next three square shapes.

_____

_____

(b) Check your predictions.

	1	2	3	4	5
1	1	2	3	4	5
2	2	4	6	8	10
3	3	6	9	12	15
4	4	8	12	16	20
5	5	10	15	20	25

	1	2	3	4	5	6
1	1	2	3	4	5	6
2	2	4	6	8	10	12
3	3	6	9	12	15	18
4	4	8	12	16	20	24
5	5	10	15	20	25	30
6	6	12	18	24	30	36

	1	2	3	4	5	6	7
1	1	2	3	4	5	6	7
2	2	4	6	8	10	12	14
3	3	6	9	12	15	18	21
4	4	8	12	16	20	24	28
5	5	10	15	20	25	30	35
6	6	12	18	24	30	36	42
7	7	14	21	28	35	42	49

Total _____   Total _____   Total _____

3 Try to find a quick way of discovering the total for the 8th, 9th and 10th squares. Check your predictions. Explain how you worked out the answer.

_____

_____

Square Number Investigations–1

A square number is made when a number is multiplied by itself. For example, 3 x 3 = 9, so 9 is a square number. See how square numbers grow.

1
(1 x 1 = 1)

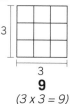
4
(2 x 2 = 4)

9
(3 x 3 = 9)

16
(4 x 4 = 16)

1 Investigate what happens when you square odd numbers and even numbers.

$1^2 =$ _____   1 $2^2 =$ _____ ☐

$3^2 =$ _____   9 $4^2 =$ _____ ☐

$5^2 =$ _____   5 $6^2 =$ _____ ☐

$7^2 =$ _____ ☐ $8^2 =$ _____ ☐

$9^2 =$ _____ ☐ $10^2 =$ _____ ☐

$11^2 =$ _____ ☐ $12^2 =$ _____ ☐

$13^2 =$ _____ ☐ $14^2 =$ _____ ☐

$15^2 =$ _____ ☐ $16^2 =$ _____ ☐

$17^2 =$ _____ ☐ $18^2 =$ _____ ☐

$19^2 =$ _____ ☐ $20^2 =$ _____ ☐

$21^2 =$ _____ ☐ $22^2 =$ _____ ☐

$23^2 =$ _____ ☐ $24^2 =$ _____ ☐

$25^2 =$ _____ $26^2 =$ _____

4^2 is read as four squared and means 4 x 4 (=16).

There is an interesting pattern to be found in the last digits.

Write the last digit of each answer in the boxes.

2 Complete the following statements.

(a) If you square an odd number, the result will be (**odd**) (**even**) .

(b) If you square an even number, the result will be (**odd**) (**even**) .

(c) What do you notice about the final digits of odd square numbers and even square numbers?

odd: _____

even: _____

Square Number Investigations–2

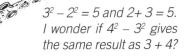

Many interesting patterns may be found within square numbers. For example:

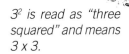 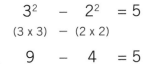 

3^2 is read as "three squared" and means 3×3.

$$3^2 \quad - \quad 2^2 \quad = 5$$
$$(3 \times 3) \quad - \quad (2 \times 2)$$
$$9 \quad - \quad 4 \quad = 5$$

$3^2 - 2^2 = 5$ and $2 + 3 = 5$. I wonder if $4^2 - 3^2$ gives the same result as $3 + 4$?

1 Try subtracting the totals of more square numbers to see if the pattern continues.

(a) $4^2 - 3^2$ → $16 - 9 = 7$ → $3 + 4 = 7$

(b) $5^2 - 4^2$ → $25 - 16 = 9$ → $4 + 5 = 9$

(c) $6^2 - 5^2$ → _____ → _____

(d) $7^2 - 6^2$ → _____ → _____

(e) $8^2 - 7^2$ → _____ → _____

(f) $9^2 - 8^2$ → _____ → _____

(g) $10^2 - 9^2$ → _____ → _____

2 Predict the results of:

(a) $20^2 - 19^2 =$ _____

(b) $25^2 - 24^2 =$ _____

(c) $100^2 - 99^2 =$ _____

(d) Explain how you worked out the answers. _____

_____

_____

3 Look at the pattern that results when you find the difference between alternate square numbers.

(a) $4^2 - 2^2 =$ _____ → _____

(b) $5^2 - 3^2 =$ _____ → _____

(c) $6^2 - 4^2 =$ _____ → _____

(d) $7^2 - 5^2 =$ _____ → _____

(e) $8^2 - 6^2 =$ _____ → _____

(f) Predict what the difference between 9^2 and 7^2 will be. _____

(g) Explain how you worked out the difference. _____

_____

_____

Patterns in Mathematics

One Up, One Down

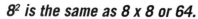

8^2 is the same as 8 x 8 or 64.

If you add one and subtract one, the multiplication created is 9 x 7.

1   (a)   Multiply nine by seven. _____ (Remember 8 x 8 = 64)

    (b)   What do you notice about the two answers? _____

_____

2   Try some others.

    (a)   12 x 12 = _____   (b)   14 x 14 = _____   (c) 20^2 = _____

        13 x 11 = _____   15 x 13 = _____   21 x 19 = _____

    (d)   What seems to be happening? _____

_____

3   Write your own set of questions that follow the same pattern.

    (a)   ____ x ____ = ____   (b)   ____ x ____ = ____   (c)   ____ x ____ = ____

        ____ x ____ = ____   ____ x ____ = ____   ____ x ____ = ____

    (d)   What happens every time? _____

_____

4   (a)   If 26^2 = 676, what does 27 x 25 produce? _____

_____

    (b)   Explain how you worked out your answer. _____

_____

5   (a)   What happens if you add two and subtract two?

        i.e. 8^2 is 8 x 8 = 64

             10 x 6 = 60

    (b)   What do you think will happen if you add three and subtract three?

_____

_____

Investigate	Test your prediction.

Triangular Numbers

1 The set of numbers shown below form a triangular pattern.
Continue the drawings to show the fifth and sixth triangular numbers.

1	2	3	4	5	6

2 (a) Write down the number of dots in each triangle.

    1   3   6   ____   ____   ____

(b) If the third triangular number is six, what is the fifth triangular number? _____

3 Continue the pattern to show the first ten triangular numbers.

1, 3, 6, ____, ____, ____, ____, ____, ____, ____.

> What does "consecutive" mean?

4 What do you notice about the triangular numbers?
Complete the boxes to show how the numbers are increasing.

+2	+3							

1   3   6 __ __ __ __ __ __ __

5 What happens when you add two consecutive triangular numbers?

(a)   1 + 3 = ____   ____ + ____ = _____

            3 + 6 = ____   ____ + ____ = _____

            6 + ____ = ____   ____ + ____ = _____

> It means one after the other, e.g. the 3rd and 4th triangular number or the 7th and 8th.

(b) What patterns do you notice? _____

6 The triangular numbers may be redrawn to look like this:

     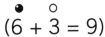

_____ _____

> Hey! That pattern looks like a square.

(a) Continue the drawing.
When two consecutive triangular numbers are joined it looks like this:

● ○ ○
● ● ○   (6 + 3 = 9)
● ● ●

The pattern 1, 4, 9, 16, ... is known as the square numbers because the dots may be drawn in the shape of a square.

(b) List the next six square numbers. 1, 4, 9, 16, _____

Counting Rectangles

How many rectangles can you find in the figure?

It may help to start with some simpler examples and build up.

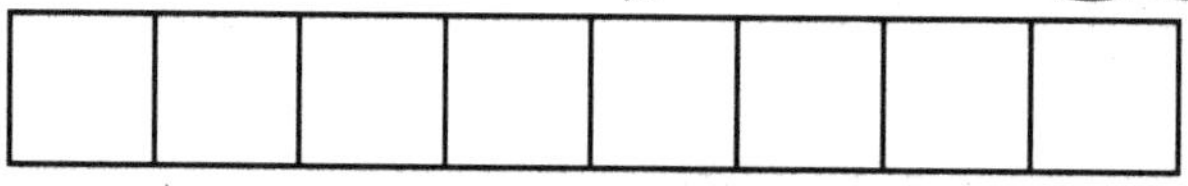

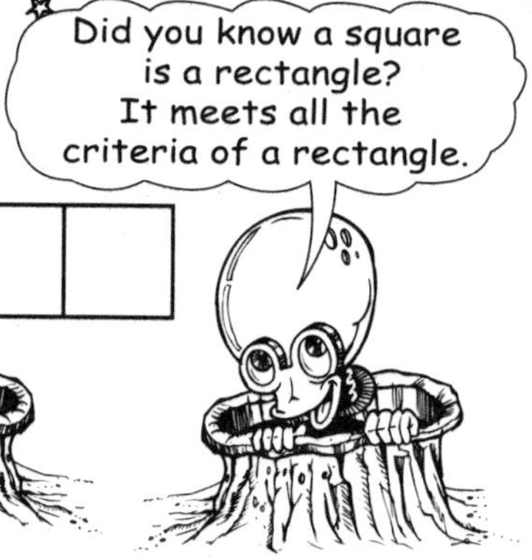

Did you know a square is a rectangle? It meets all the criteria of a rectangle.

1 Continue the table until you reach the figure above.

Figures	Squares	Rectangles	Total
 3 rectangles!	1	0	1
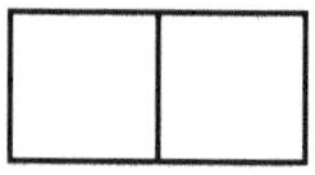	2	1	3
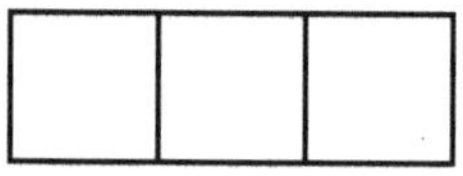	3	3	6
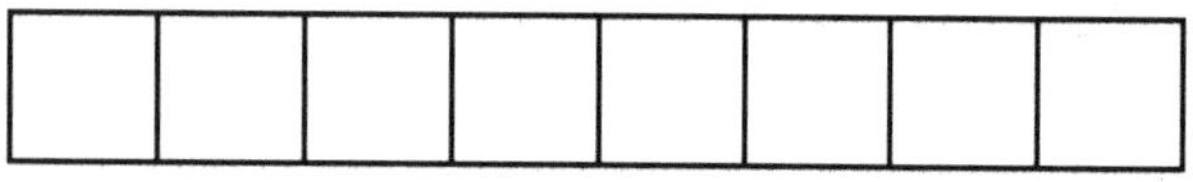			

2 (a) What patterns do you notice? _____

_____

I think I recognize these numbers ...

(b) How many rectangles, in total, were there in the original figure?

_____

Intersections

An intersection is the place where two lines meet.

The maximum number of intersections produced from two straight lines is one.

The maximum number of intersections produced from three straight lines is three.

1 What is the maximum number of intersections that may be produced from:

(a) four lines? _____

(b) five lines? _____

Before you write your answer, try drawing the lines with as many intersections as you can on the back of the sheet.

2 Place your data in a table.

(a)

Number of Lines	2	3	4	5
Maximum Number of Intersections	1	3		

(b) Do you recognize this pattern? These are the _____ numbers.

3 Predict how many intersections there will be for:

six lines _____ seven lines _____ ten lines _____

Staircases

A staircase is to be built using blocks.

I recognize this pattern!

1   **2**   **3**   **4**   **5**

The first step requires one block. Two more blocks are required to build step two and so on.

4 (a) Draw the fifth step on the dotted line above.

Predict the tenth step.

(b) How many blocks are used? _____

(c) Continue the table.

Number of Steps	1	2	3	4	5	6	7	10
Total Number of Blocks	1	3	6					

Square Numbers and Triangular Numbers

Triangular numbers and square numbers are related. If you add two consecutive triangular numbers you end up with a square number.

1 Complete the following chart, showing how the triangular numbers are related to the square numbers.

▲ Number	▲ Number	■ Square Number	Total
3	6	9	3 + 6 = 9
6	10	16	6 + 10 = 16
10	15	25	
15	21		15 + 21 = ____
21		49	
		64	
		81	
		100	

2 Sometimes real-life situations lead to interesting number patterns. Consider the following question. How many handshakes would eight people make if they all shook hands with one another?

Use the back of this sheet for your diagrams.

I find a diagram helps me to think.

This pattern looks familiar.

Work with a partner to solve this puzzle.

_____ handshakes.

(You can test your answer at lunchtime.)

© Didax Educational Resources® – www.didax.com

Pentagonal Numbers

A pentagon is a five-sided shape. When dots are arranged into a pentagonal shape a pattern is formed.

I thought it was a building!

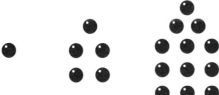

There is a building called the Pentagon. It has five sides.

1. (a) Draw the next two pentagonal numbers above.

   (b) List the first five pentagonal numbers by adding the dots.

   _____, _____, _____, _____, _____

   (c) Name the next two pentagonal numbers. _____, _____
   Use the back of this sheet for your diagram. Can you see a pattern?

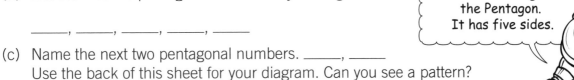

1   5   12   22   35...

(+4) (+7) (+10) (+13)

2. A pentagonal number is really made up of a square number and a triangular number.

   Draw the third and fourth pentagonal numbers showing the square number part and the triangular number part. (Use ■ for the square part and ▲ for the triangular part.)

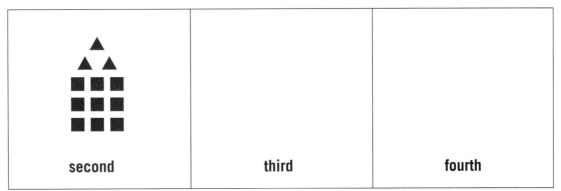

| second | third | fourth |

3. Calculate the triangular and square numbers that make up the following pentagonal numbers.

   (a) 51 = _____ + _____
              ■   ▲

   (b) 70 = _____ + _____
              ■   ▲

   (c) 92 = _____ + _____
              ■   ▲

I wonder if there are hexagonal numbers?

Triangular numbers, square numbers and pentagonal numbers are part of a family of numbers called "figurative numbers" because the patterns look like geometric shapes.

   (d) 117 = _____ + _____
                ■   ▲

   (e) 145 = _____ + _____
                ■   ▲

Patterns in Mathematics

An ISBN, or International Standard Book Number, is found on most books. An ISBN helps to identify a book.

Here is an example of an ISBN.

0 646 15584 9   *or*

`0 646 15584 9`

The ISBN is made up of ten digits divided into four parts.

- *The first number identifies the language in which the book is written. A zero or one at the start tells us that the book was written in English. Two indicates French, three German and so on.*

- *The second group of numbers identifies the publisher.*

- *The third group represents the book title and type of binding. The final number is the "check digit."*

The check digit is designed to identify any mistakes in the ISBN. To calculate the check digit, the first nine numbers are used. To find out if the check digit is correct, follow this procedure. (You will need to multiply, add, divide and subtract!)

0	x	10	=	0
6	x	9	=	54
4	x	8	=	32
6	x	7	=	42
1	x	6	=	6
5	x	5	=	25
5	x	4	=	20
8	x	3	=	24
4	x	2	=	8

Total = `211`

211   + 9 = 220

9 is the final number of the ISBN (the check digit).

 ✔

1 Check if the ISBN of the book in the picture is correct.

Step 1. Write the first 9 digits of the ISBN in each box in order.

Step 2. Multiply the first number by 10, the second number by 9, the third by eight and so on.

Step 3. Add the nine answers from Step 2.

Step 4. To find the check digit, calculate how much needs to be added to the total to make the next multiple of eleven. This should give the check digit, (8).

`1 863 11583 8`

Let me see … the multiples of 11 are 11, 22, 33, 44, 55, 66, 77, 88, 99, 110, 121, 132, 143, 154, 165, 176, 187, 198, 209, 220, 231 and so on. The closest multiple of 11 after 211 is 220.

Step 1   **Step 2**

☐ x 10 = _____

☐ x 9 = _____

☐ x 8 = _____

☐ x 7 = _____

☐ x 6 = _____

☐ x 5 = _____

☐ x 4 = _____

☐ x 3 = _____

☐ x 2 = _____

Step 3

Total = ☐

Step 4

_____ + ☐ = _____

Is the ISBN correct?

✫ (**Yes**) ✫   (**No**)

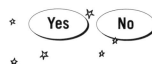

ISBN-2

1 Follow the steps to check these ISBNs.

(a) **1 854 52022 9**

☐ x 10 = _____
☐ x 9 = _____
☐ x 8 = _____
☐ x 7 = _____
☐ x 6 = _____
☐ x 5 = _____
☐ x 4 = _____
☐ x 3 = _____
☐ x 2 = _____
_____ + ☐ = _____
Total + ? = **Next multiple of eleven**
ISBN correct? Ⓨ Ⓝ

(b) **0 471 29563 9**

☐ x 10 = _____
☐ x 9 = _____
☐ x 8 = _____
☐ x 7 = _____
☐ x 6 = _____
☐ x 5 = _____
☐ x 4 = _____
☐ x 3 = _____
☐ x 2 = _____
_____ + ☐ = _____
Total + ? = **Next multiple of eleven**
ISBN correct? Ⓨ Ⓝ

(c) **0 946 99403 X**

☐ x 10 = _____
☐ x 9 = _____
☐ x 8 = _____
☐ x 7 = _____
☐ x 6 = _____
☐ x 5 = _____
☐ x 4 = _____
☐ x 3 = _____
☐ x 2 = _____
_____ + ☐ = _____
Total + ? = **Next multiple of eleven**
ISBN correct? Ⓨ Ⓝ

(d) **0 862 27650 0**

☐ x 10 = _____
☐ x 9 = _____
☐ x 8 = _____
☐ x 7 = _____
☐ x 6 = _____
☐ x 5 = _____
☐ x 4 = _____
☐ x 3 = _____
☐ x 2 = _____
_____ + ☐ = _____
Total + ? = **Next multiple of eleven**
ISBN correct? Ⓨ Ⓝ

(e) Explain what it means when an ISBN ends in:

(i) **X** _____

_____

_____

(ii) **0** _____

_____

_____

2 On the back of this sheet, check the ISBNs below.

(a) 1 823 17492 9

(b) 1 862 82062 7

(c) 0 853 21147 9

(d) 0 863 13584 6

3 Choose four books of your own. Write their titles and ISBNs on a separate sheet of paper. Check them by following the steps above.

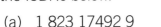

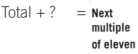
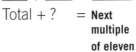

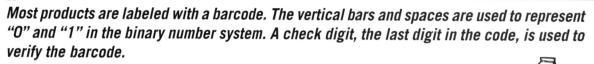

Barcodes–1

Most products are labeled with a barcode. The vertical bars and spaces are used to represent "0" and "1" in the binary number system. A check digit, the last digit in the code, is used to verify the barcode.

To check a barcode, follow these steps.

Step 1   *Write down the barcode number.*

Second last digit ⎯⎯   Check digit

9   ③ 1 ① 6 ⑨ 7   ① 0 ① 5 ⑧   ③⎯

Step 2   *Draw a box around the last digit (the check digit).*

Step 3   *Draw a circle around the second to last digit and then every second one from there on.*

Step 4   *Add all the numbers in circles. (3 + 1 + 9 + 1 + 1 + 8 = 23)*

Step 5   *Multiply the result by 3. (23 x 3 = 69)*

Step 6   *Add all the digits without a circle or box around them. (9 + 1 + 6 + 7 + 0 + 5 = 28)*

Step 7   *Add the results from steps 5 and 6. (69 + 28 = 97)*

Step 8   *Add the check digit (3) to the total (97) and you should get a multiple of ten. If not, you have either made a mistake, or the barcode is printed incorrectly.*

---

1   Check the following barcodes.

(a)

9 310055 149809

(b)

9 300657 015336

3 + 0 + 5 + 1 + 9 + 0 = ☐

☐ x 3 = _____

_____ = ☐

_____ + _____ = ☐

☐ + 9 = _____

3 + 0 + 5 + 0 + 5 + 3 = ☐

☐ x 3 = _____

_____ = ☐

_____ + _____ = ☐

☐ + 6 = _____

Is the answer a multiple of ten? (Y)(N)

Is the answer a multiple of ten? (Y)(N)

2   Check the following barcodes. Use the back of this sheet for your calculations.

(a)

9 300657 304461

(b)

9 300652 010756

(c)

9 310060 006234

3   Find more barcodes and check them. Hint: Barcodes are found on most food products.

Barcodes–2

1 Find four barcodes. Glue or copy them into the boxes. Use the steps to check if they are correct.

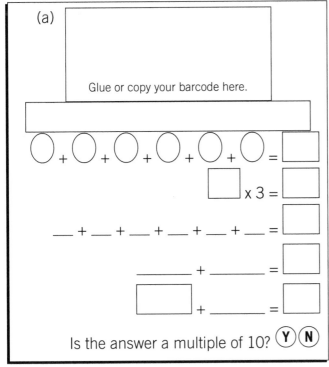

(a)

Glue or copy your barcode here.

$\bigcirc + \bigcirc + \bigcirc + \bigcirc + \bigcirc + \bigcirc = \square$

$\square \times 3 = \square$

$\_\_ + \_\_ + \_\_ + \_\_ + \_\_ + \_\_ = \square$

$\_\_\_\_\_ + \_\_\_\_\_ = \square$

$\square + \_\_\_\_\_ = \square$

Is the answer a multiple of 10? Ⓨ Ⓝ

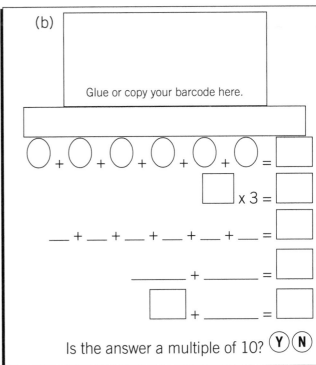

(b)

Glue or copy your barcode here.

$\bigcirc + \bigcirc + \bigcirc + \bigcirc + \bigcirc + \bigcirc = \square$

$\square \times 3 = \square$

$\_\_ + \_\_ + \_\_ + \_\_ + \_\_ + \_\_ = \square$

$\_\_\_\_\_ + \_\_\_\_\_ = \square$

$\square + \_\_\_\_\_ = \square$

Is the answer a multiple of 10? Ⓨ Ⓝ

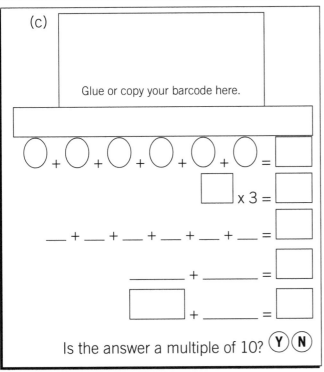

(c)

Glue or copy your barcode here.

$\bigcirc + \bigcirc + \bigcirc + \bigcirc + \bigcirc + \bigcirc = \square$

$\square \times 3 = \square$

$\_\_ + \_\_ + \_\_ + \_\_ + \_\_ + \_\_ = \square$

$\_\_\_\_\_ + \_\_\_\_\_ = \square$

$\square + \_\_\_\_\_ = \square$

Is the answer a multiple of 10? Ⓨ Ⓝ

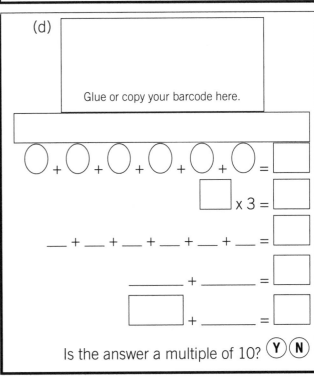

(d)

Glue or copy your barcode here.

$\bigcirc + \bigcirc + \bigcirc + \bigcirc + \bigcirc + \bigcirc = \square$

$\square \times 3 = \square$

$\_\_ + \_\_ + \_\_ + \_\_ + \_\_ + \_\_ = \square$

$\_\_\_\_\_ + \_\_\_\_\_ = \square$

$\square + \_\_\_\_\_ = \square$

Is the answer a multiple of 10? Ⓨ Ⓝ

Discuss these questions with a friend.

2 (a) How are barcodes used in supermarkets?

(b) What does the checkout person do if the barcode doesn't scan properly?

(c) Discontinued food is often marked with a thick marker crossing out the barcode. Why do you think this is?

Patterns in Mathematics

Pascal's Triangle–1

The number pattern below is named after a French mathematician, Blaise Pascal, who lived in the 17th Century. When the numbers are placed in a triangular formation, the pattern becomes clearer.

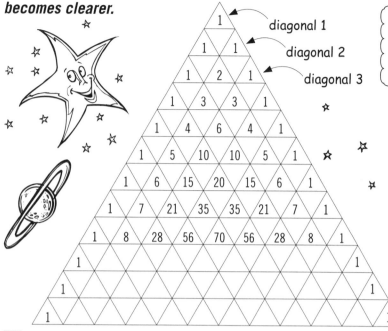

diagonal 1
diagonal 2
diagonal 3

Even though the triangle is named after Pascal, it has been in existence from around 200 B.C.!

1. Can you see how each row is related to the previous row? Add three more rows on to the triangle.

2. Look at the numbers along each of the diagonals marked. What do you notice?

Diagonal 1: _____

Diagonal 2: _____

Diagonal 3: _____

---

3. (a) Calculate the value of:

$11^2 = (11 \times 11) =$ _____

$11^3 = (11 \times 11 \times 11) =$ _____

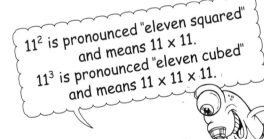

11^2 is pronounced "eleven squared" and means 11 × 11.
11^3 is pronounced "eleven cubed" and means 11 × 11 × 11.

(b) Look at the third and fourth row of Pascal's triangle. What do you notice?

_____

_____

(c) Predict the value of 11^4 (11 × 11 × 11 × 11). _____

Check your prediction with a calculator. Were you correct? (**Yes**) (**No**)

(d) What happens with 11^5? _____

_____

Pascal's Triangle–2

Each number in Pascal's triangle is found by adding the two numbers above it.

1. Complete the triangle.

2. Calculate the total for each of the first **five** rows only.

Total

	(1)
	(1 + 1)
	(1 + 2 + 1)
	(1 + 3 + 3 + 1)
	(1 + 4 + 6 + 4 + 1)

3. Predict the total for the next three rows

6th row _____ 7th row _____ 8th row _____

4. Check your predictions by adding the numbers in each of the 6th, 7th and 8th rows.

(a) Were you correct?   (**Yes**)   (**No**)

(b) What do you notice? _____

_____

5. (a) Add the numbers in the third (highlighted) diagonal as far as 55. _____

(b) Compare your answer with the total of the number that is one diagonal across and one row down.

What do you notice? _____

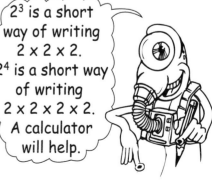

challenge

Complete the following table for the powers of two.

2^0	2^1	2^2	2^3	2^4	2^5	2^6	2^7	2^8	2^9	2^{10}
1	2	4								

Describe any relationships you can see between the powers of two and the rows on Pascal's triangle.

2^3 is a short way of writing 2 x 2 x 2. 2^4 is a short way of writing 2 x 2 x 2 x 2. A calculator will help.

_____

_____

Coloring Pascal's Triangle–1

Each number in Pascal's triangle is found by adding the two numbers above it.
For example:

```
   1   2   1
 1   3   3   1
1   4   6   4   1
```

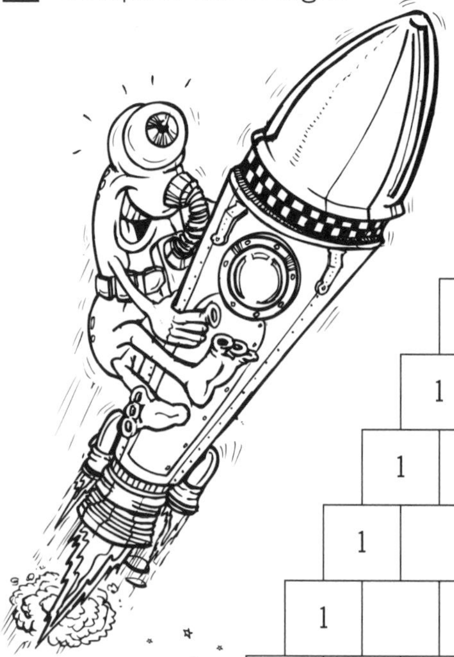

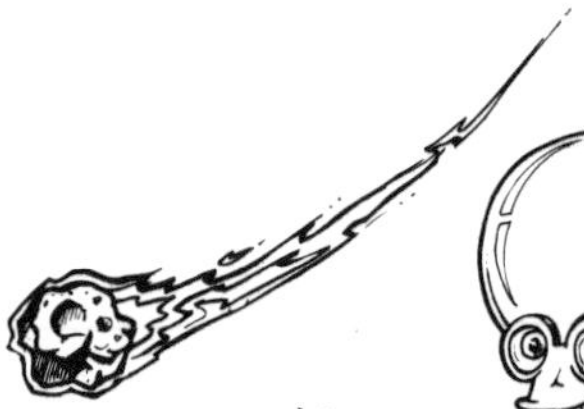

1 Complete the triangle.

> Can you find the symmetry in the triangle that makes it easier to complete?

Triangle rows:

```
                        1
                      1   1
                    1   2   1
                  1   3   3   1
                1   4   6   4   1
              1               1
            1                   1
          1                       1
        1                           1
      1                               1
    1       120 210 252 210             1
  1                                       1
1   12  66  220  495  792  924  792         1
  1                                       1
1           3,003 3,432 3,003 2,002 1,001 364  91  14  1
  1                                         1
1  16  120  560  1,820  4,368  8,008  11,440  12,870  11,440  8,008                 1
```

2 Color the multiples of five; i.e. 5, 10, 15, 20, etc.

3 Write about the pattern that you notice. _____

_____

_____

Coloring Pascal's Triangle–2

Each number in Pascal's triangle is found by adding the two numbers above it.

For example:

1 Complete the triangle.

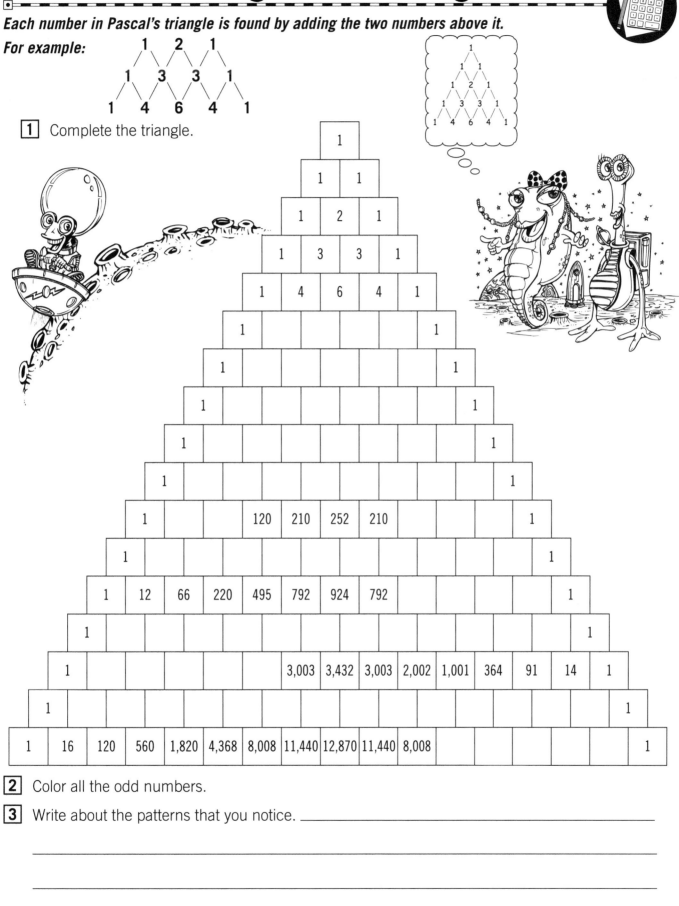

2 Color all the odd numbers.

3 Write about the patterns that you notice. _____

_____

_____

Fun with Fibonacci

Leonardo Fibonacci, an Italian mathematician, discovered the following number sequence;

1, 1, 2, 3, 5, 8, 13, 21, 34 etc.

1 Look for a pattern to see how each new number in the sequence is formed. Explain how to continue the sequence.

_____

_____

> I think it has to do with adding.

2 Write the next three numbers in the sequence. _____ _____ _____

3 Some interesting patterns come to light when you divide one Fibonacci number by the one before it. Complete the following calculations. Continue using the Fibonacci sequence in (i) and (j).

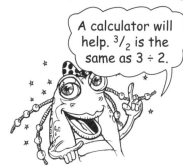

> A calculator will help. $^3/_2$ is the same as $3 \div 2$.

(a) $^1/_1$ = _____   (f) $^{13}/_8$ = _____

(b) $^2/_1$ = _____   (g) $^{21}/_{13}$ = _____

(c) $^3/_2$ = _____   (h) $^{34}/_{21}$ = _____

(d) $^5/_3$ = _____   (i) _____ = _____

(e) $^8/_5$ = _____   (j) _____ = _____

4 What do notice about the answers? _____

_____

5 (a) • Choose any three consecutive Fibonacci numbers, e.g. 2, 3, 5.

   • Square the middle number, e.g. 3^2 (3 x 3) = 9.

   • Multiply the two numbers left, e.g. 2 x 5 = _____.

   • What do you notice? _____

(b) Try some other consecutive Fibonacci numbers, e.g. 5, 8, 13; 8, 13, 21; 13, 21, 34. What do you notice?

_____

---

challenge

> The ratio 1:1.6 is known as the Golden Ratio and may be found in architecture. The ancient Greeks believed that certain rectangles were more pleasing to the eye than others.

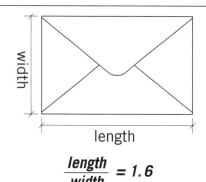

width

length

$$\frac{length}{width} = 1.6$$

Measure the length and width of a standard envelope and divide the length by the width.

What do you notice?

_____

_____

_____

Fibonacci Facts

Leonardo Fibonacci was an Italian mathematician. He was also known as Leonardo the son of Bonacci and as Leonardo da Pisa.

He wrote a book called "LiberAbachi" (The Book of the Abacus) published in 1202. In the book, this sequence is found:

1, 1, 2, 3, 5, 8, 13 …

Each number in the sequence is formed by adding the previous two numbers in the sequence.

1 List the first twelve terms of the sequence.

_____

2 (a) Add the first eight Fibonacci numbers and compare it to the tenth Fibonacci number.

What do you notice? _____

_____

(b) Now add the first nine and then the first ten Fibonacci numbers and compare their totals to the eleventh and twelfth Fibonacci numbers.

What do you notice? _____

_____

3 Now try adding every second Fibonacci term—i.e. 1 + 2 + 5 …—and note any relationships.

_____

_____

4 What happens if you add every second term, beginning with 1, 3, 8 …? _____

_____

5

The Fibonacci sequence was made famous by a mathematician called Lucas.

Lucas created his own Fibonacci sequence but he started with the numbers 1 and 3. The sequence is now called the Lucas sequence.

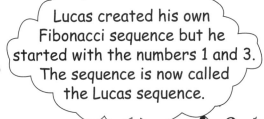

The Lucas sequence is:

1, 3, 4, 7, 11, _____, _____, _____, _____, _____, _____

Find out how the sequence is formed and add the next six numbers.

Golden Rectangles

1 Measure the three items listed below to find out whether they are Golden Rectangles. Look around your classroom to find seven more rectangle shapes to measure.

Item	Length of longer side	Length of shorter side	Longer side / Shorter side	Golden Rectangle
Envelope				Yes/No
Playing card				Yes/No
Book cover				Yes/No
				Yes/No
				Yes/No
				Yes/No
				Yes/No
				Yes/No
				Yes/No
				Yes/No

2 Explain what you discovered. _____

_____

_____

3 The Golden Rectangle was often used in art and in architecture, e.g. the Parthenon in Greece. Use the Internet and other resources to find out where else the Golden Rectangle occurs. Summarize your findings below.

The Golden Ratio

1 Measure the length and width of the rectangle shown at the bottom of the page.

length = _____ mm

width = _____ mm

divide the length by the width

_____ ÷ _____ = _____

What do you notice about the results?

_____

_____

2 Cut a square from the rectangle (100 mm x 100 mm).

(a) Measure the piece that is left.

_____ mm x _____ mm

(b) Divide the length by the width.

_____ mm ÷ _____ mm = _____ mm

What do you notice? _____

_____

_____

3 Try starting with a different Golden Rectangle. You may like to try a rectangle that is 80 mm x 50 mm. Keep measuring and cutting squares and rectangles.
Describe your findings. _____

_____

_____

_____

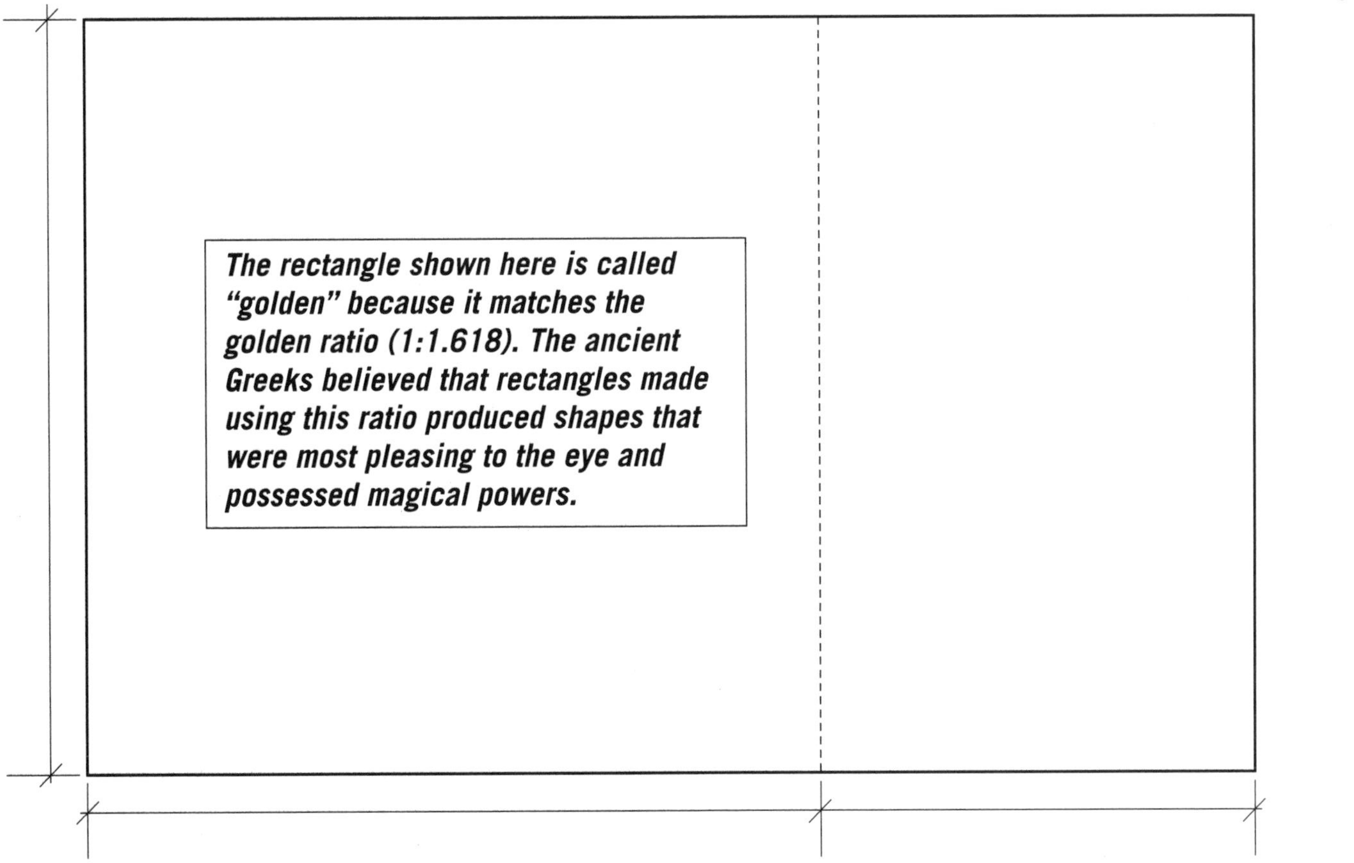

(c) Cut a square (60mm x 60mm) from your piece of paper and measure the rectangle that remains. Measure the length and width. What do you notice when the length is divided by the width?

_____

_____

_____

✂

The rectangle shown here is called "golden" because it matches the golden ratio (1:1.618). The ancient Greeks believed that rectangles made using this ratio produced shapes that were most pleasing to the eye and possessed magical powers.

Prime Time

Eratosthenes was a Greek mathematician who lived over 2,000 years ago. He came up with a system for finding "Prime Numbers."

1 Follow the steps below to find all the prime numbers between 1 and 100.

Step 1. Cross out 1.

Step 2. Circle 2.

Step 3. Cross out all the multiples of 2.

Step 4. Circle 3 and cross out all the multiples of 3.

Step 5. Four should already be crossed out, so circle 5 and cross out all the multiples of 5.

Step 6. Six will already be crossed out, so circle 7 and cross out all the multiples of 7.

Step 7. Circle all the remaining numbers. These numbers are called "prime numbers."

1	2	3	4	5	6	7
8	9	10	11	12	13	14
15	16	17	18	19	20	21
22	23	24	25	26	27	28
29	30	31	32	33	34	35
36	37	38	39	40	41	42
43	44	45	46	47	48	49
50	51	52	53	54	55	56
57	58	59	60	61	62	63
64	65	66	67	68	69	70
71	72	73	74	75	76	77
78	79	80	81	82	83	84
85	86	87	88	89	90	91
92	93	94	95	96	97	98
99	100	101	102	103	104	105

2 What pattern did you notice with the circled prime numbers? _____

_____

3 Look at the multiples of 6 and note where all the prime numbers (except 2 and 3) lie.

_____

4 An 18th century German mathematician, called Goldbach, made the following conjecture:

> *Every even number, except two, is the sum of two prime numbers.*

(a) Choose even numbers and test Goldbach's statement below.

_____

_____

_____

(b) How many different ways can you make 20 using prime numbers?

_____ ways.

(c) How many different ways can you make 42 using prime numbers?

_____ ways.

Number Patterns–1

Complete the sequences by adding the next three numbers. Describe the rule for forming the sequence, e.g. adding three.

1

(a) 4, 8, 12, 16, _____, _____, _____

_____

_____

(b) 5, 9, 13, 17, _____, _____, _____

_____

_____

(c) 91, 87, 83, 79, _____, _____, _____

_____

_____

(d) How are the three sequences similar?

_____

(e) How are the three sequences different?

_____

2

(a) 1, 3, 9, 27, _____, _____, _____

_____

_____

(b) 3, 6, 12, 24, _____, _____, _____

_____

_____

(c) 1, 10, 100, 1,000, _____, _____, _____

_____

_____

(d) How are the three sequences similar?

_____

(e) How are the three sequences different?

_____

3

(a) 1, 4, 10, 22, _____, _____, _____

_____

(b) 1, 3, 7, 15, _____, _____, _____

_____

(c) 2, 3, 5, 9, _____, _____, _____

_____

(d) How are the three sequences similar?

_____

(e) How are the three sequences different?

> The rule for these has two parts!

4 Create your own number pattern on a separate page for a friend to solve.

Number Patterns–2

Continue the numbers for each sequence. Explain how each sequence is produced.

1 5, 9, 13, 17, _____, _____, _____

_____

_____

2 73, 67, 61, 55, _____, _____, _____

_____

_____

3 4, 7, 11, 16, _____, _____, _____

_____

_____

4 2, 5, 10, 17, _____, _____, _____

_____

_____

5 2, 4, 8, 16, _____, _____, _____

_____

_____

6 15, 30, 45, 60, _____, _____, _____

_____

_____

7 96, 48, 24, 12, _____, _____, _____

_____

_____

8 1, 3, 9, 27, _____, _____, _____

_____

_____

9 1, 4, 9, 16, 25, _____, _____, _____

_____

_____

10 0.5, 0.6, 0.7, 0.8, _____, _____, _____

_____

_____

11 1 000, 100, 10, 1, _____, _____, _____

_____

12 1, 1, 2, 3, 5, 8, _____, _____, _____

_____

13 1, 2, 4, _____, _____, _____

_____

14 1, 2, 4, _____, _____, _____

_____

Note: It takes more than three numbers to create a sequence!

16 Write three sequences of your own. Give them to a friend to solve,

(a) _____, _____, _____, _____, _____

_____

(b) _____, _____, _____, _____, _____

_____

(c) _____, _____, _____, _____, _____

_____

Try to finish this sequence in a different way to the previous one.

Super Sequences

1 (a) Continue the following sequence.

1, 2, 3, 4, 1, 2, 3, 4, _____, _____, _____, _____, _____, _____, _____

(b) Predict the 20th number in this sequence. _____

(c) Explain how you made your prediction. _____

_____

(c) Predict the 39th number in this sequence. _____

(d) Explain how you made your prediction. _____

_____

(e) If you continued this sequence so that there were 43 numbers, how many 2s would it have?

_____

(f) Explain how you arrived at your answer. _____

_____

2 *All the numbers from 1 to 100 are to be made using stickers for each of the digits. For example, to make the number 79 would require two stickers – a sticker with 7 on it and another with 9 on it.*

| 7 | 9 |

(a) How many stickers with 7 showing on them would you buy? _____

(b) Explain how you arrived at your answer. _____

_____

_____

(c) How many of each digit should you order? _____

_____

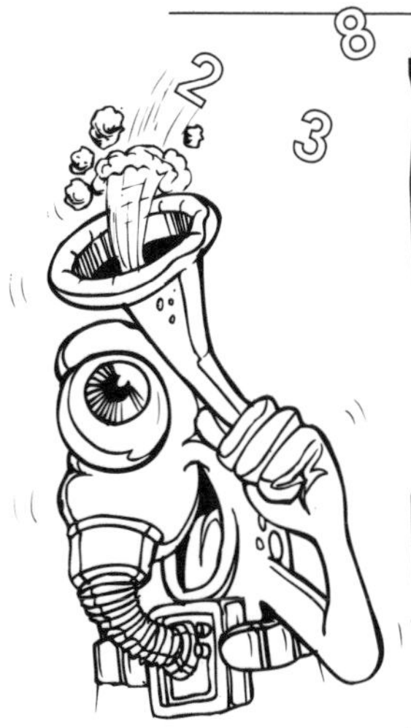

Use the space below for your investigations. A 1–100 grid will help!

Power Patterns

Do you know about squares of numbers and cubes of numbers? For example, 6^2 is read as "six squared" or "six raised to the power of two" and means that you would multiply six by itself; i.e. 6 x 6.

6^3 would be read as "six cubed" or "six raised to the power of three" and means that you would multiply six by itself and by itself again; i.e. 6 x 6 x 6.

1 (a) Complete the following.

Most calculators have a square button that looks like x^2 and some calculators have a power button that looks like y^x.

6 x 6 x 6 x 6 ...

$6^0 = 1$

$6^1 = 6$

$6^2 \rightarrow 6 \times 6 = 36$

$6^3 \rightarrow 6 \times 6 \times 6 =$ _____

$6^4 \rightarrow 6 \times 6 \times 6 \times 6 =$ _____

$6^5 \rightarrow$ _____ = _____

$6^6 \rightarrow$ _____ = _____

$6^7 \rightarrow$ _____ = _____

$6^8 \rightarrow$ _____ = _____

$6^9 \rightarrow$ _____ = _____

(b) What do you notice about the last digit in each answer? _____

(c) Predict what the final digit for 6^{10} will be? _____ Check your answer with a calculator.

2 (a) Investigate the powers of 2, 3 and 4. Record your results in the table below.

(b) Describe the patterns you see.

(i) Powers of 2 _____

(ii) Powers of 3 _____

(ii) Powers of 4 _____

2^x	Result	Last Digit	3^x	Result	Last Digit	4^x	Result	Last Digit
2^1	2	2	3^1	3	3	4^1	4	4
2^2	4	4	3^2	9	9	4^2		
2^3	8	8	3^3	27	7	4^3		
2^4	16	6	3^4	81	1	4^4		
2^5	32	2	3^5			4^5		
2^6	64	4	3^6			4^6		
2^7			3^7			4^7		
2^8			3^8			4^8		
2^9			3^9			4^9		
2^{10}			3^{10}			4^{10}		

A Million Dollars Per Month

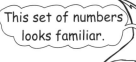

Would you prefer to be paid $1,000,000 for one month (30 days) or 1¢ for the first day of the month, 2¢ for the second day, 4¢ for the third day, 8¢ for the fourth day and so on?

This set of numbers looks familiar.

1 Which form of payment would you prefer? Explain your choice.

_____

_____

2 Complete the table below to investigate the second pay option.

Day	Payment in Cents	Total payment in Cents	Day	Payment in Cents	Total payment in Cents
1	1	1	16		
2	2	3	17		
3	4	7	18		
4	8	15	19		
5	16	31	20		
6	32	63	21		
7			22		
8			23		
9			24		
10			25		
11			26		
12			27		
13			28		
14			29		
15			30		

3 How much would you earn with the second pay option? _____

The Game of Chess

According to legend, the game of chess was invented for an Indian king. The king was so happy with the game that he offered the inventor a reward. The inventor simply asked for a few grains of wheat—or so it seemed! He asked that one grain of wheat be placed on the first square of the chessboard, two grains on the second square, four grains on the third square, eight grains on the fourth square and so on. The king thought it was a simple request.

- Investigate the request on a separate piece of paper. A calculator may help, although not for long.

Ordered Pairs

A pattern may be written in several different ways. One way involves writing pairs of numbers. Find the missing numbers from each set of ordered pairs. Explain the rule you used to find the missing numbers. For example,

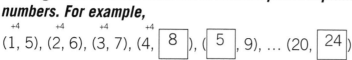

(1, 5), (2, 6), (3, 7), (4, $\boxed{8}$), ($\boxed{5}$, 9), ... (20, $\boxed{24}$)

Adding four pattern _____

> Pairs of numbers are usually coordinates that are used in graphs or games (like battleships).

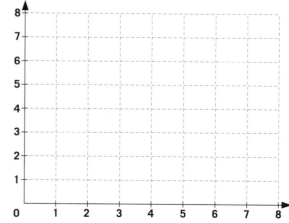

1 (a) (20, 18), (19, 17), (18, 16), (17, ☐), (☐, 14), ... (10, ☐)

_____

(b) (1, 3), (2, 6), (3, 9), (4, ☐), (☐, 15), ... (22, ☐)

_____

(c) (40, 30), (39, 29), (38, 28), (37, ☐), (36, ☐), ... (10, ☐)

_____

(d) (1, 1), (2, 4), (3, 9), (4, 16), (5, ☐), (6, ☐), ... (20, ☐)

_____

(e) (1, 9), (2, 18), (3, 27), (4, ☐), (5, ☐), ... (9, ☐)

_____

(f) (7, 23), (8, 24), (9, 25), (10, ☐), (11, ☐), ... (75, ☐)

_____

(g) (42, 39), (41, 38), (40, 37), (39, ☐), (38, ☐), ... (3, ☐)

_____

2 Ordered pairs may be used to describe a set of points to be plotted onto a graph.

(a) Plot the following points onto the graph: (0, 1), (1, 2), (2, 3), (3, 4), (4, 5), (5, 6)

(b) Describe what you see.

_____

_____

_____

(c) Continue the line a little further and list the points it passes through.

_____

© Didax Educational Resources® – www.didax.com

Paths and Patterns

A network is made up of paths, vertices and regions. An example of a network is shown. This network has four paths, three vertices and three regions (two inside the network and the region around the network).
Network: Regions (R) = 3, Vertices (V) = 3, Paths or Edges (E) = 4

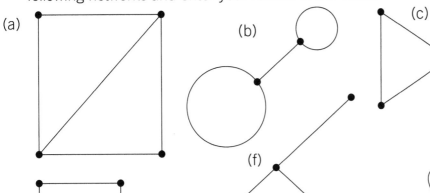

1 Count the number of paths, vertices and regions for each of the following networks and enter your results in the table.

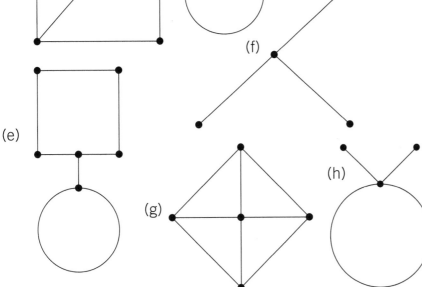

A mathematician named Euler discovered a relationship between the regions, vertices and edges. He added the regions and vertices and found a rule that linked them to the number of edges.

Shape	Regions (R)	Vertices (V)	Paths or Edges (E)	R + V – E
(a)				
(b)				
(c)				
(d)				
(e)				
(f)				
(g)				
(h)				

2 Do you notice a pattern in the table? Explain below. _____

_____

3 Complete the rule: **R + V – E =** ▢

Circle Patterns

The following activity involves joining points on a nine-point circle to produce nonagons. Starting at the point labelled 9, join every second point in the circle in a clockwise direction until you get back to the starting point.

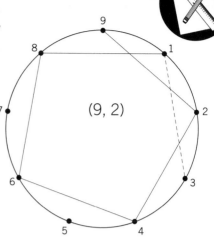

(9, 2)

☐1 Color in the middle shape to highlight the nonagon (nine-sided polygon). The whole figure is called a "star polygon."

☐2 Now draw a (9, 7) star polygon by joining every seventh point, starting at 9.

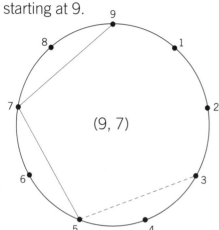

(9, 7)

☐3 What do you notice about the (9, 2) and the (9, 7) star polygons?

_____

_____

_____

☐4 Draw a (9, 3) and a (9, 6) star polygon below.

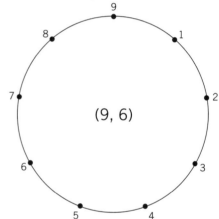

(9, 3)   (9, 6)

☐5 (a) Predict what will happen when you draw a (9, 4) and a (9, 5) star polygon.

_____

_____

(b) Try them. Are you correct? Ⓨ Ⓝ

(9, 4)   (9, 5)

Ten-point Circles

The following activity involves joining points on a ten-point circle to produce polygons.

1 Starting at the point labelled 10, join every second point in the circle in a clockwise direction until you get back to the starting point.

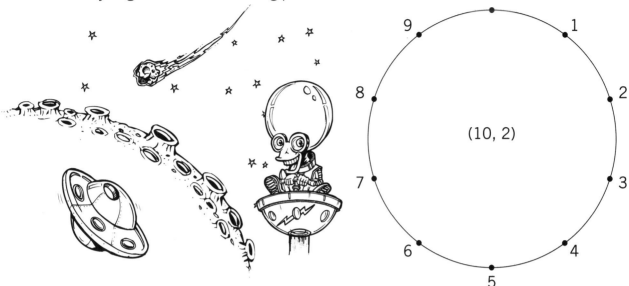

(10, 2)

2 Investigate other ten-point circles. (10, 8) means start at ten and join every 8th point in the circle in a clockwise direction until you get back to the starting point.

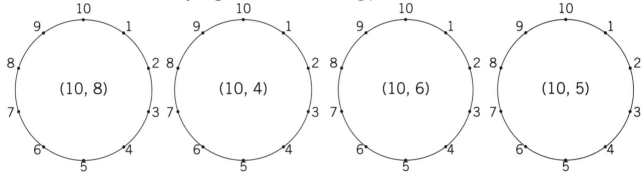

(10, 8)   (10, 4)   (10, 6)   (10, 5)

3 Choose your own ten-point circles to draw.

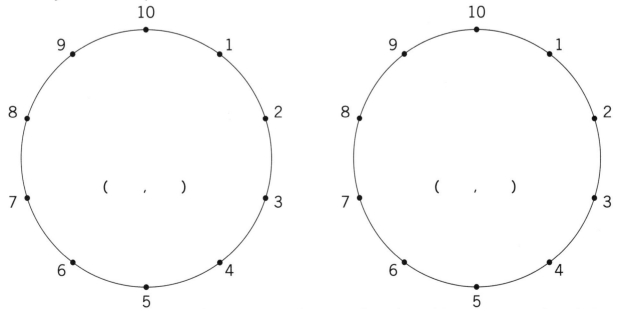

(,)   (,)

4 Investigate other patterns on different types of circles. Try using odd and even numbered circles and look for patterns.

Joining Function Machines

Appendix 1

(See activities – pages 24 – 26)

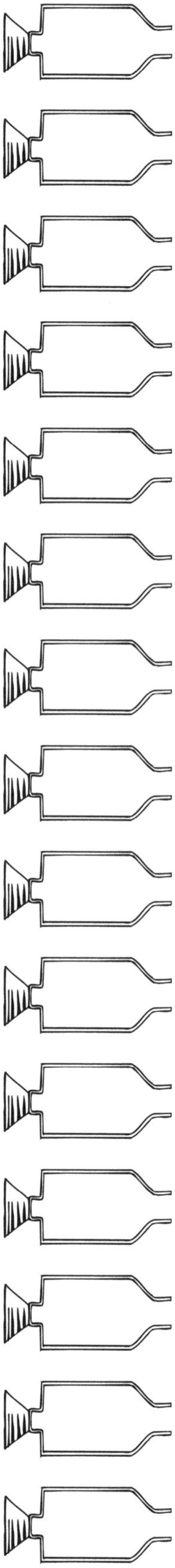

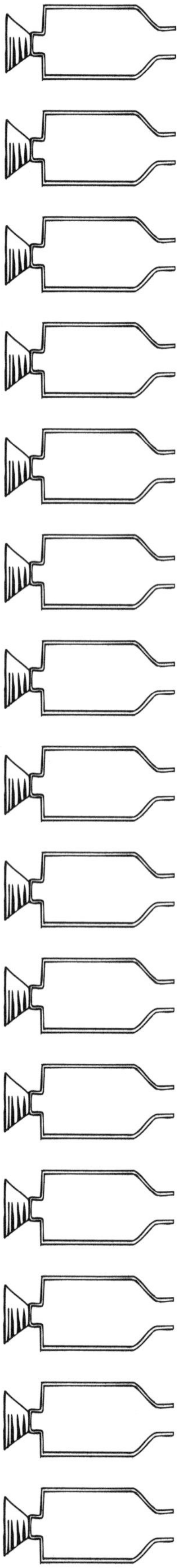

(a) ______________ ______________ ______

(b) ______________ ______________ ______

(c) ______________ ______________ ______

(d) ______________ ______________ ______

(e) ______________ ______________ ______

(f) ______________ ______________ ______

(g) ______________ ______________ ______

(h) ______________ ______________ ______

(i) ______________ ______________ ______

(j) ______________ ______________ ______

(k) ______________ ______________ ______

(l) ______________ ______________ ______

(m) ______________ ______________ ______

(n) ______________ ______________ ______

(o) ______________ ______________ ______

Multiplication Grids

(See activity pages 33 – 35)

(a)

X	1	2	3	4	5	6	7	8	9
1	1	2	3	4	5	6	7	8	9
2	2	4	6	8	10	12	14	16	18
3	3	6	9	12	15	18	21	24	27
4	4	8	12	16	20	24	28	32	36
5	5	10	15	20	25	30	35	40	45
6	6	12	18	24	30	36	42	48	54
7	7	14	21	28	35	42	49	56	63
8	8	16	24	32	40	48	56	64	72
9	9	18	27	36	45	54	63	72	81

(d)

X	1	2	3	4	5	6	7	8	9
1	1	2	3	4	5	6	7	8	9
2	2	4	6	8	10	12	14	16	18
3	3	6	9	12	15	18	21	24	27
4	4	8	12	16	20	24	28	32	36
5	5	10	15	20	25	30	35	40	45
6	6	12	18	24	30	36	42	48	54
7	7	14	21	28	35	42	49	56	63
8	8	16	24	32	40	48	56	64	72
9	9	18	27	36	45	54	63	72	81

(b)

X	1	2	3	4	5	6	7	8	9
1	1	2	3	4	5	6	7	8	9
2	2	4	6	8	10	12	14	16	18
3	3	6	9	12	15	18	21	24	27
4	4	8	12	16	20	24	28	32	36
5	5	10	15	20	25	30	35	40	45
6	6	12	18	24	30	36	42	48	54
7	7	14	21	28	35	42	49	56	63
8	8	16	24	32	40	48	56	64	72
9	9	18	27	36	45	54	63	72	81

(e)

X	1	2	3	4	5	6	7	8	9
1	1	2	3	4	5	6	7	8	9
2	2	4	6	8	10	12	14	16	18
3	3	6	9	12	15	18	21	24	27
4	4	8	12	16	20	24	28	32	36
5	5	10	15	20	25	30	35	40	45
6	6	12	18	24	30	36	42	48	54
7	7	14	21	28	35	42	49	56	63
8	8	16	24	32	40	48	56	64	72
9	9	18	27	36	45	54	63	72	81

(c)

X	1	2	3	4	5	6	7	8	9
1	1	2	3	4	5	6	7	8	9
2	2	4	6	8	10	12	14	16	18
3	3	6	9	12	15	18	21	24	27
4	4	8	12	16	20	24	28	32	36
5	5	10	15	20	25	30	35	40	45
6	6	12	18	24	30	36	42	48	54
7	7	14	21	28	35	42	49	56	63
8	8	16	24	32	40	48	56	64	72
9	9	18	27	36	45	54	63	72	81

(f)

X	1	2	3	4	5	6	7	8	9
1	1	2	3	4	5	6	7	8	9
2	2	4	6	8	10	12	14	16	18
3	3	6	9	12	15	18	21	24	27
4	4	8	12	16	20	24	28	32	36
5	5	10	15	20	25	30	35	40	45
6	6	12	18	24	30	36	42	48	54
7	7	14	21	28	35	42	49	56	63
8	8	16	24	32	40	48	56	64	72
9	9	18	27	36	45	54	63	72	81

Grid Paper

(See activity page 38)

(a)

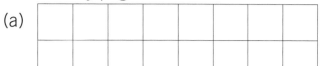

(d)

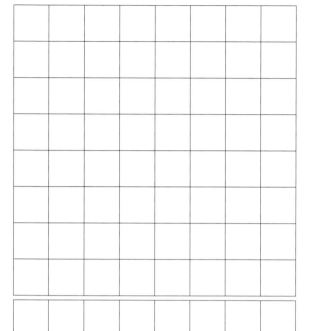

(b)

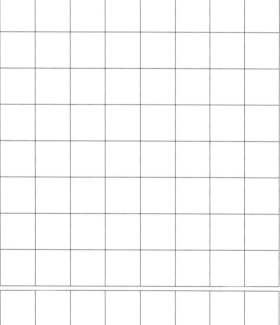

(e)

(c)

(f)

Addition Patterns .... page 8

1.

+	0	1	2	3	4	5	6	7	8	9
0	0	1	2	3	4	5	6	7	8	9
1	1	2	3	4	5	6	7	8	9	10
2	2	3	4	5	6	7	8	9	10	11
3	3	4	5	6	7	8	9	10	11	12
4	4	5	6	7	8	9	10	11	12	13
5	5	6	7	8	9	10	11	12	13	14
6	6	7	8	9	10	11	12	13	14	15
7	7	8	9	10	11	12	13	14	15	16
8	8	9	10	11	12	13	14	15	16	17
9	9	10	11	12	13	14	15	16	17	18

2. (a) 45 55 65 75
   (b) The totals for each row increase by 10.
3. (a) (i) 85 (ii) 95 (iii) 105 (iv) 115 (v) 125
4. (a) 145 (b) 165 (c) 45

More Addition Patterns .... page 9

1.

+	0	1	2	3	4	5	6	7	8	9
0	0	1	2	3	4	5	6	7	8	9
1	1	2	3	4	5	6	7	8	9	10
2	2	3	4	5	6	7	8	9	10	11
3	3	4	5	6	7	8	9	10	11	12
4	4	5	6	7	8	9	10	11	12	13
5	5	6	7	8	9	10	11	12	13	14
6	6	7	8	9	10	11	12	13	14	15
7	7	8	9	10	11	12	13	14	15	16
8	8	9	10	11	12	13	14	15	16	17
9	9	10	11	12	13	14	15	16	17	18

2. (a) 9 18 27 36 45
   (b) The answers are the multiples of nine.

More Addition Patterns (continued) .... page 10

3. (a) 63 72 81 90
   (b) Teacher check
4. (a) The pattern would reverse; i.e. 81, 72, 63 because of the symmetry in the addition table.
   (b) 81
5. (a) 0, 2, 6, 12, 20, 30, 42, 56, 72, 90, 90, 88, 84, 78, 70, 60, 48, 34, 18
   (b) Initially the sequence increases in the following pattern 2, 4, 6, 8, 10 . . . and then decreases by 2, 4, 6 . . .

Patterns Within the Addition Table .... page 11

1. (a) – (d) Teacher check
   (e) All the answers divide by four without leaving a remainder.
2. (a) Teacher check
   (b) In each box, one number will be repeated. Multiply this number by four to find the total.
3. (a) Teacher check
   (b) Multiply the middle number by nine.

 = 900

More Patterns in the Addition Table .... page 12

1. (a) 6 x 10 = 60 (b) 8 x 8 = 64 (c) 64 – 60 = 4
2. (a) – (b) Teacher check (c) The difference is always 4.
3. (b) The difference is always 9.

Place Value Patterns - 1 .... page 13

1. (a) 26   (e) 2.6
   (b) 260   (f) 0.26
   (c) 2,600   (g) 0.026
   (d) 26,000   (h) 0.0026
2. (a) There are several patterns but basically the numbers are increasing or decreasing by a factor of ten.
   (b) The digits 2 and 6 are the same.
   (c) The place value of the digits changes.
3. (a) The value of the number increases.
   (b) The value of the number decreases.
4. Consider the effect on place value of multiplying or dividing by ten.
5. (a) 390   (b) 39,000   (c) 0.39

Place Value Patterns - 2 .... page 14

1. (a) 43   (f) 430
   (b) 4.3   (g) 4,300
   (c) 0.43   (h) 43,000
   (d) 0.043   (i) 430,000
   (e) 0.0043   (j) 4,300,000
2. (a) The numbers are decreasing by a factor of ten.
       The numbers are increasing by a factor of ten.
   (b) Consider the effect of place value when dividing by powers of ten, e.g. 100, 0.01
3. (a) 7.9   (d) 790
   (b) 0.079   (e) 7,900
   (c) 0.0079   (f) 790,000

Place Value Patterns - 3 .... page 15

1. (a) 4,671   (e) 467.1
   (b) 467.1   (f) 46.71
   (c) 46.71   (g) 46.71
   (d) 4.671   (h) 4.671
2. (a) The digits remain the same in each case but the value of the number changes. The size of the number varies by a factor of ten depending on the placement of the decimal point.
   (b) The answers all contain the digits 4, 6, 7 and 1.
   (c) The place value of the digits changes.
3. (a) 379.2   (e) 37,920
   (b) 37.92   (f) 37,920
   (c) 37.92   (g) 3,792
   (d) 37.92   (h) 37.92
4. The digits remain the same but by noticing the effect of powers of ten on the size of the number, you can estimate the relative size of the number.

Noticing Nines - 1 .... page 16

1. (a)
   9   9 = 9
   18   1 + 8 = 9
   27   2 + 7 = 9
   36   3 + 6 = 9
   45   4 + 5 = 9
   54   5 + 4 = 9
   63   6 + 3 = 9
   72   7 + 2 = 9
   81   8 + 1 = 9
   90   9 + 0 = 9

   (b) There are several patterns. For example, the tens digit goes up by one and the units digit goes down by one each time. When you add the tens and units digits, nine is always the answer.

2. When subtracting the tens and ones digits the following pattern is formed: 9, 7, 5, 3, 1, 3, 5, 7, 9.

Answers

Noticing Nines - 2 .... page 17

1. Teacher check
   Odd, even, odd, even, etc. Answers alternate.
2. (a) Adding digits along the diagonals produces eight each time.
   (b) Adding digits along the other diagonal produces ten each time.
3. Subtracting the digits produces a 6, 4, 2, 0, 2, 4, 6 pattern along both diagonals.
4.
6,624	$6+6+2+4$	$= 18,$ $1 + 8 = 9$
3,933	$3+9+3+3$	$= 18$
5,535	$5+5+3+5$	$= 18$
3,024	$3+0+2+4$	$= 9$
16,533	$1+6+5+3+3$	$= 18,$ $1 + 8 = 9$
5. The digits always add to nine.

The 99 Times Table .... page 18

1.
1 x 99 = 99	8 x 99 = 792	15 x 99 = 1,485
2 x 99 = 198	9 x 99 = 891	16 x 99 = 1,584
3 x 99 = 297	10 x 99 = 990	17 x 99 = 1,683
4 x 99 = 396	11 x 99 = 1,089	18 x 99 = 1,782
5 x 99 = 495	12 x 99 = 1,188	19 x 99 = 1,881
6 x 99 = 594	13 x 99 = 1,287	20 x 99 = 1,980
7 x 99 = 693	14 x 99 = 1,386	21 x 99 = 2,079
2. (b) The units decrease, the tens are ten nines, ten eights, ten sevens, etc.; the hundreds digits increase.
   (c) Adding the digits of the answers will always produce a multiple of nine.
3. A number is divisible by nine (without leaving a remainder) if the digits add to nine.

Eleven Times .... page 19

1. (a) 121   (b) 132   (c) 143   (d) 154   (e) 165
   (f) 176   (g) 187
2. Teacher check
3. (a) 198   (b) 209
   (c) The pattern changes when you reach 20 x 11. (20 x 11 = 220)
4. (a) 561   (b) 374   (c) 297   (d) 473   (e) 341
   (f) 792
5. To multiply 68 by 11 you apply the same process: Separate the 6 and 8, with the units digit being 8. 6 + 8 = 14; therefore, the tens digit will be 4 and the hundreds digit will become 7 instead of 6. You could also multiply by ten and then add 68.
6. (a) 649   (b) 902   (c) 847   (d) 1 023

Puzzling Patterns - 1 .... page 20

1. (a) 1,188 2,277 3,366 4,455 5,544 6,633 7,722
   (b) 9 98 987 9,876 98,765 123456 x 8 + 6 9,876,543
   (c) 9,801 98,901 989,901 9,899,901
   (d) 8,712 87,912 879,912 8,799,912 87,999,912
   (e) 11 111 1,111 11,111 111,111 123,456 x 9 + 7 11,111,111 111,111,111
   (f) 88 888 8,888 888,888 987,654 x 9 + 2 88,888,888

Puzzling Patterns - 2 .... page 21

1. (a) 111 222 333 444 555 666
   (b) 111,111 222,222 333,333 444,444 555,555 666,666
   (c) 1,287 2,376 3,465 4,554 5,643 7,821
   (d) 2,189 3,289 4,389 5,489 6,589 9,889
   (e) 25 225 625 1,225 2,025 3,025
   (f) 2385 3,285 4,185 5,085 5,985 8,685

Puzzling Patterns - 3 .... page 22

1. (a) 111,111 222,222 333,333 444,444 555,555 666,666 777,777
   (b) 9,109 18,218 27,327 36,436 45,545 54,654 81,981
   (c) 0.090909 0.181818 0.272727 0.363636 454545 0.545454 0.636363 0.727272
   (d) 0.1111111 0.2222222 0.3333333 0.4444444 0.5555555 0.6666666 0.7777777 0.8888888
   ★ 0.353535 0.123123

Staggering Additions .... page 23

1. (odd numbers)
2. (difference of 2)
3. (multiples of 3), (difference of 3)
4. (difference of 4)
5. There will be a difference of 5.
6. Teacher check
   ★ Answers will vary depending on how the rows are aligned and the numbers of rows used.

Function Machines - 1 .... page 24

1. (a) 17, 18, 19, Adds ten, ten more
   (b) 101, 102, 103, 104, Adds one, one more
   (c) 9, 10, 11, 12, 13 One less, subtract one
   (d) 10, 15, 20, 25, 30 Halve, divide in half, divide by two
   (e) 2, 3, 4, 5, 6 Divide by ten, one tenth of
   (f) 12, 24, 36, 48, 60 Multiplying by four, quadrupling
   (g) 15, 16, 17, 18, 19 Subtracting five
   (h) 5, 6, 7, 8, 9 Doubling, multiplying by two
2. Work backwards. For example, by adding 5 to the numbers coming out of the −5 function machine, you can work out which numbers entered the machine. Dividing by two will undo multiplying by two.
3. (a) 6, 7, 8, 9, 10

Function Machines - 2 .... page 25

1. (a) 5, 6, 7, 8, 9 dividing by two or halving
   (b) 5, 6, 7, 8, 9 halving – same as dividing by two
   (c) 100, 200, 300, 400 multiplying by ten
   (d) 10, 20, 30, 40, 50 dividing by ten, finding one tenth of
   (e) 10, 20, 30, 40, 50 doubling, multiplying by two
2. ÷2 and multiplying by $\frac{1}{2}$ produce the same result. ÷10 undoes x10. To work out the numbers coming into the x2 function machine, you need to halve the numbers coming out or divide the numbers coming out by two.
3. (a) 8, 9, 10
   (b) 15, 16, 17, 18, 19 then 20, 21, 22, 23, 24
   (c) The +2 and +3 function machines may be replaced by a single +5 machine. The +5 +5 machines can be replaced by a single +10 machine.
4. (a) Two subtraction machines may be replaced by a single function machine that combines the numbers from each. For example, a −2 and −4 machine may be replaced by a single −6 machine.
   (b) When an addition and subtraction machine are combined they may be replaced by a single machine. For example, a +2 and −5 machine may be replaced by a single −3 machine.

Function Machines - Finding the Program ... page 26

1. (a) +3 The machine adds three to the numbers coming through.
   (b) +10 The machine adds ten

Answers

(c) x3 Multiplies by three or triples

(d) ÷2 or $^1/_2$ Halving

(e) −5 Subtracting five

(f) x10 Multiplying by ten

(g) ÷10 or x $^1/_{10}$ Dividing by ten or multiplying by a tenth

(h) +1.5 Adding one and a half

(i) −2.5 Subtracting two and a half

(j) ÷3 or x $^1/_3$ Dividing by three or multiplying by a third

Swap the Numbers .... page 27

1. (a) Teacher check
   (b) A multiple of nine is always produced.
2. (a) 81, 72, 63, 54, 45, 36, 27, 18, 9
   (b) The answers are all multiples of nine. Basically, the 9 times table in ascending order is produced.
3. (a) 80, 81, 82, 83, 84, 85, 86, 87
   (b) Answers decrease by one.

   ★ Seventies – 70, 71, 72, 73, 74, 75, 76
   Sixties – 60, 61, 62, 63, 64, 65
   Fifties – 50, 51, 52, 53, 54

Graphing Multiples - 1 .... page 28

1. Teacher check
2. (a) 3, 6, 9, 12, 15, 18, 21, 24, 27, 30
   (b) Teacher check
   (c) The slope of the graph is steeper.
3. (a) The slope is even steeper.
   (b) 4, 8, 12, 16, 20, 24, 28, 32, 36, 40
   (c) Teacher check
4. The slope of the graph would be steeper than the graph for the multiples of two and three.

Graphing Multiples - 2 .... page 29

1. (a) 24, 30, 36, 42, 48, 54   (b) Teacher check
2. (a) 14, 21, 28, 35, 42, 49, 56   (b) Teacher check
3. The multiples of seven produce the steepest slope.
4. (a) The graph of the multiples of four would have a slope less than the multiples of seven but greater than the multiples of two.
   (b) Teacher check
5. (a) The slope of the one times table would be less than the two times table.
   (b) Teacher check   (c) Teacher check

Marking Multiples - 1 .... page 30

1. (a) Teacher check
   (b) A diagonal pattern sloping from the right to the left is produced.
2. (a) Teacher check
   (b) A diagonal pattern sloping from the left to the right is produced.
3. (a) 7 and 9   (b) Teacher check
4. (a) – (b) Teacher check
   (c) The grid would be four columns wide.

Marking Multiples - 2 .... page 31

1. (a) Teacher check   (b) Two columns are shown
2. (a) Multiples of 4 and 8
   (b) Teacher check
3. (a) 6, 12, 18, 24, 30, 36, 42, 48, 54, 60

(b) – (c) Teacher check

(d) The grid would be twelve columns wide

4. (a) Teacher check
   (b) The multiples of ten
   (c) 14, 28, 42, 56, . . .

Diagonal Dilemmas .... page 32

1. (a)

x	1	2	3	4	5	6	7	8	9
1	1	2	3	4	5	6	7	8	9
2	2	4	6	8	10	12	14	16	18
3	3	6	9	12	15	18	21	24	27
4	4	8	12	16	20	24	28	32	36
5	5	10	15	20	25	30	35	40	45
6	6	12	18	24	30	36	42	48	54
7	7	14	21	28	35	42	49	56	63
8	8	16	24	32	40	48	56	64	72
9	9	18	27	36	45	54	63	72	81

   (b) Teacher check
   (c) 2, 1, 0, 1, 2, 3, 4, 5
   (d) The pattern decreases by ones and then increases by ones.
2. (a) Teacher check
   (b) 6, 5, 4, 3, 2, 1, 0, 1
   (c) The pattern decreases by ones and then increases by ones.
3. (a) - (b) Teacher check

   ★ 5, 6, 7, 8, 9, 10 . . . The pattern increases by ones.

Multiplication Patterns .... page 33

1. (a) – (b) Teacher check
   (c) The same result is produced.
2. (a) Teacher check
   (b) When the pairs of numbers in opposite corners of a 3 x 3 rectangle are multiplied, the same result is produced.
3. (a) – (b) Teacher check. The same result occurs with rectangles of any size.
4. Teacher check

In the Middle .... page 34

1. (a) 54   (b) 9
   (c) 126 ÷ 14 = 9,   225 ÷ 25 = 9,   144 ÷ 16 = 9,   576 ÷ 64 = 9
   (d) The answer is always nine.
2. The total will be the center number multiplied by nine.
   (a) – (d) Teacher check

The Reverse "L" .... page 35

1. 1, 8, 27, 64
2. 1, 8, 27, 64
3. They are the same.
4. (a) Teacher check
   (b) Teacher check
   (c) 125 (5 x 5 x 5), 216 (6 x 6 x 6), 343 (7 x 7 x 7)
5. Cubic numbers grow very quickly.

Answers

Robots - 1 .... page 36

1. (a)
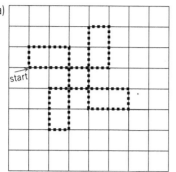

   (b) Pattern is repeated four times.

2. (a) (b)
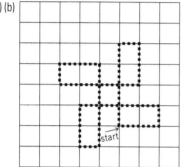

   (c) Pattern is repeated four times.

3. Teacher check

Robots - 2 .... page 37

1. (a) Teacher check
   (b) Four rectangles
       are produced.
       You finish at
       the starting
       point.
   (c) Teacher check

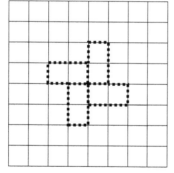

   Note: Any set of instructions in the order 1, 2, 3 or 3, 1, 2 will produce the same pattern. As long as the order is preserved, the pattern will be the same.
2. Teacher check

Even More Robots .... page 38

1. (a) – (c) Teacher check
   (d) All three paths
       will be the same.
   (e) The order of steps
       remains the same.

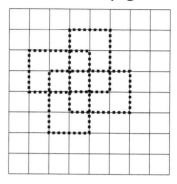

2. (a) A 3-step, 2-step,
       4-step robot walk
   (b) The pattern is
       flipped over.

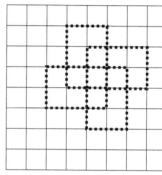

3. (a) A 2-step, 3-step,
       6-step robot walk
   (b) Note:
       the square hole in
       the middle of the
       pattern. A similar
       pattern is produced
       when a 3-step,
       6-step, 2-step
       pattern is drawn.

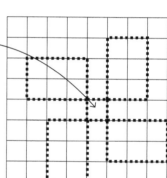

Answers

4. (a) Note that this robot walk also produces a square hole in the middle.

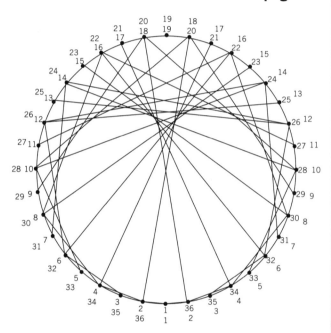

(b) If the difference between the two short sides and the long side is 1, a square hole 1 x 1 is produced. If the difference is 2, a 2 x 2 square hole is produced. No difference = no hole.

Circle Patterns .... page 39

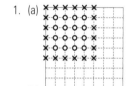

Digit Sums .... page 40

1. (a) 4 + 1 fi 5

   (b) 3 + 8 + 2 fi 13 fi 1 + 3 = 4

   (c) 4 + 8 + 8 + 6 → 26 → 2 + 6 = 8

2. (a)

	Digit Sum		Digit Sum
1 x 6 = 6	6	11 x 6 = 66	3
2 x 6 = 12	3	12 x 6 = 72	9
3 x 6 = 18	9	13 x 6 = 78	6
4 x 6 = 24	6	14 x 6 = 84	3
5 x 6 = 30	3	15 x 6 = 90	9
6 x 6 = 36	9	16 x 6 = 96	6
7 x 6 = 42	6	17 x 6 = 102	3
8 x 6 = 48	3	18 x 6 = 108	9
9 x 6 = 54	9	19 x 6 = 114	6
10 x 6 = 60	6	20 x 6 = 120	3

(b) A repeating 6, 3, 9 pattern occurs.

(c) 9, 6, 6

3.

2x	Digit Sum	3x	Digit Sum	4x	Digit Sum	5x	Digit Sum	7x	Digit Sum	8x	Digit Sum	9x	Digit Sum
2	2	3	3	4	4	5	5	7	7	8	8	9	9
4	4	6	6	8	8	10	1	14	5	16	7	18	9
6	6	9	9	12	3	15	6	21	3	24	6	27	9
8	8	12	3	16	7	20	2	28	1	32	5	36	9
10	1	15	6	20	2	25	7	35	8	40	4	45	9
12	3	18	9	24	6	30	3	42	6	48	3	54	9
14	5	21	3	28	1	35	8	49	4	56	2	63	9
16	7	24	6	32	5	40	4	56	2	64	1	72	9
18	9	27	9	36	9	45	9	63	9	72	9	81	9
20	2	30	3	40	4	50	5	70	7	80	8	90	9
22	4	33	6	44	8	55	1	77	5	88	7	99	9
24	6	36	9	48	3	60	6	84	3	96	6	108	9
26	8	39	3	52	7	65	2	91	1	104	5	117	9
28	1	42	6	56	2	70	7	98	8	112	4	126	9
30	3	45	9	60	6	75	3	105	6	120	3	135	9

Adding Consecutive Numbers .... page 41

1. (a) 4, 9, 16, 25, 36, 49, 64, 81, 100 – all square numbers.

   (b)

No. of Odd Numbers	Consecutive Odd Numbers	Total
1	1	1
2	1 + 3	4
3	1 + 3 + 5	9
4	1 + 3 + 5 + 7	16
5	1 + 3 + 5 + 7 + 9	25
6	1 + 3 + 5 + 7 + 9 + 11	36
7	1 + 3 + 5 + 7 + 9 + 11 + 13	49
8	1 + 3 + 5 + 7 + 9 + 11 + 13 + 15	64
9	1 + 3 + 5 + 7 + 9 + 11 + 13 + 15 + 17	81
10	1 + 3 + 5 + 7 + 9 + 11 + 13 + 15 + 17 + 19	100

2. (a) 20^2 → (20 x 20) = 400

   (b) $1^2 = 1$, $2^2 = 4$, $3^2 = 9$, $4^2 = 16$, etc. So it stands to reason that adding the first twenty consecutive odd numbers would be 20^2 or 400.

Visual Patterns .... page 42

1. (a)

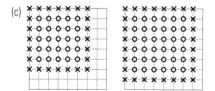

(b)

Pegs on the outside	8	12	16	20	24	28
Holes in the middle	1	4	9	16	25	36

(c)

(d) Holes are all square numbers.

2. (a) 81

   (b) Take four from the number of pegs on the outside, then divide the result by four and square the answer; that will indicate the number of holes in the middle.

2. (a) The set of triangular numbers is formed.

Answers

Multiplication Squares .... page 43

1. (a) Teacher check   (b) 1, 9, 36, 100

   (c) They are all square numbers.

2. (a) Teacher check

   A "skipping" pattern is formed.

   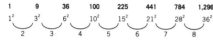

   (b) 225, 441, 784, 1,296

3. 64, 81, 100 Squaring the length of the square. For example, width of eighth square is 8, 8 x 8 = 64.

Square Number Investigations - 1 .... page 44

1.

	Last Digit		Last Digit
$1^2 = 1$	1	$2^2 = 4$	4
$3^2 = 9$	9	$4^2 = 16$	6
$5^2 = 25$	5	$6^2 = 36$	6
$7^2 = 49$	9	$8^2 = 64$	4
$9^2 = 81$	1	$10^2 = 100$	0
$11^2 = 121$	1	$12^2 = 144$	4
$13^2 = 169$	9	$14^2 = 196$	6
$15^2 = 225$	5	$16^2 = 256$	6
$17^2 = 289$	9	$18^2 = 324$	4
$19^2 = 361$	1	$20^2 = 400$	0
$21^2 = 441$	1	$22^2 = 484$	4
$23^2 = 529$	9	$24^2 = 576$	6
$25^2 = 625$	5	$26^2 = 676$	6

2. (a) Squaring an odd number produces an odd result.

   (b) Squaring an even number produces an even result.

   (c) A "1, 9, 5, 9, 1" pattern continues in the units digit of the squares of odd numbers.

   A "4, 6, 6, 4, 0" pattern continues in the units digit of the squares of even numbers.

Square Number Investigations - 2 .... page 45

1. (a) $4^2 - 3^2 = 7$ $3 + 4 = 7$

   (b) $5^2 - 4^2 = 9$ $4 + 5 = 9$

   (c) $6^2 - 5^2 = 11$ $5 + 6 = 11$

   (d) $7^2 - 6^2 = 13$ $6 + 7 = 13$

   (e) $8^2 - 7^2 = 15$ $7 + 8 = 15$

   (f) $9^2 - 8^2 = 17$ $8 + 9 = 17$

   (g) $10^2 - 9^2 = 19$ $9 + 10 = 19$

2. (a) $20^2 - 19^2 = 39$ $20 + 19 = 39$

   (b) $25^2 - 24^2 = 49$ $25 + 24 = 49$

   (c) $100^2 - 99^2 = 199$ $100 + 99 = 199$

3. (a) 12   (b) 16   (c) 20

   (d) 24   (e) 28   (f) 32

   (g) The difference always goes up by four.

Note: If you add the two numbers involved and double, you will find the resulting difference.

i.e. $6^2 - 4^2 \rightarrow 36 - 16 = 20$

$6 + 4 = 10 \rightarrow 2 \times 10 = 20$

One Up, One Down .... page 46

1. (a) 9 x 7 = 63

   (b) 64 and 63 are one apart.

2. (a) 144, 143

   (b) 196, 195

   (c) 400, 399

   (d) There is always a difference of one.

3. (a)-(c) Teacher check

   (d) A difference of one is produced in every case.

4. (a) 675

   (b) Subtract one from 676.

5. (a) The difference is always four.

   (b) The difference is always nine.

Triangular Numbers .... page 47

1.

2. (a) 10, 15, 21

   (b) 15

3. 10, 15, 21, 28, 36, 45, 55

4. +4, +5, +6, +7, +8, +9, +10

5. 1 + 3 = 4, 3 + 6 = 9, 6 + 10 = 16, 10 + 15 = 25

   15 + 21 = 36, 21 + 28 = 49, 36 + 45 = 81, 45 + 55 = 100

   The set of square numbers is formed.

6. (a)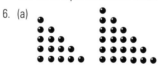

   (b) 25, 36, 49, 64, 81, 100

Counting Rectangles .... page 48

1.

Figures	Squares	Rectangles	Total
☐	1	0	1
☐☐	2	1	3
☐☐☐	3	3	6
☐☐☐☐	4	6	10
☐☐☐☐☐	5	10	15
☐☐☐☐☐☐	6	15	21
☐☐☐☐☐☐☐	7	21	28
☐☐☐☐☐☐☐☐	8	28	36

2. (a) The set of triangular numbers is formed.

(b) There were 36 rectangles in the original figure.

Intersections .... page 49

1. (a) 6 intersections

(b) 10 intersections

2. (a)

Number of Lines	2	3	4	5
Maximum Number of Intersections	1	3	6	10

(b) triangular

3. Six lines – 15 intersections

Seven lines – 21 intersections

Eight lines – 28 intersections

Note: The formula $\frac{n(n-1)}{}$ represents the number of intersections for "n" lines.

Square Numbers and

Triangular Numbers .... page 50

1.

$3+6 = 9$	$21+28 = 49$
$6+10 = 16$	$28+36 = 64$
$10+15 = 25$	$36+45 = 81$
$15+21 = 36$	$45+55 = 100$

2. 28 handshakes

Note:

People	2	3	4	5	6	7	8
Handshakes	1	3	6	10	15	21	28

The triangular number pattern is formed.

The relationship of people to handshakes is $\frac{n(n-1)}{}$.

Pentagonal Numbers .... page 51

1. (a)

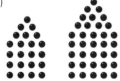

(b) 1, 5, 12, 22, 35

(c) 35, 51

Pattern No.	1	2	3	4	5	6	7	8	9
Pentagonal No.	1	5	12	22	35	51	70	92	117

4 7 10 13 16 19 22 25
  3   3   3   3   3   3   3

2. Teacher check

3. (a) $51 = \frac{36}{\blacksquare} + \frac{15}{\blacktriangle}$ (b)

(c) $92 = \frac{64}{\blacksquare} + \frac{28}{\blacktriangle}$ (d)

$70 = \frac{49}{\blacksquare} + \frac{21}{\blacktriangle}$

$117 = \frac{81}{\blacksquare} + \frac{36}{\blacktriangle}$

(e) $145 = \frac{100}{\blacksquare} + \frac{45}{\blacktriangle}$

ISBN - 1 .... page 52

1. $(1 \times 10) + (8 \times 9) + (6 \times 8) + (3 \times 7) + (1 \times 6) + (1 \times 5) + (5 \times 4) + (8 \times 3) + (3 \times 2) = 212$

$212 + 8 = 220$ (next multiple of 11)

ISBN - 2 .... page 53

1. (a) $(1 \times 10) + (8 \times 9) + (5 \times 8) + (4 \times 7) + (5 \times 6) + (2 \times 5) + (0 \times 4) + (2 \times 3) + (2 \times 2) = 200$

$200 + 9 = 209$ (next multiple of 11)

(b) $(0 \times 10) + (4 \times 9) + (7 \times 8) + (1 \times 7) + (2 \times 6) + (9 \times 5) + (5 \times 4) + (6 \times 3) + (3 \times 2) = 200$

$200 + 9 = 209$

(c) $(0 \times 10) + (9 \times 9) + (4 \times 8) + (6 \times 7) + (9 \times 6) + (9 \times 5) + (4 \times 4) + (0 \times 3) + (3 \times 2) = 276$

$276 + 10 = 286$

(d) $(0 \times 10) + (8 \times 9) + (6 \times 8) + (2 \times 7) + (2 \times 6) + (7 \times 5) + (6 \times 4) + (5 \times 3) + (0 \times 2) = 220$

220 is a multiple of 11 therefore the check digit is zero.

(e) X is the Roman numeral for 10.

0 means the digits add to a multiple of 11

2. (a) $(1 \times 10) + (8 \times 9) + (2 \times 8) + (3 \times 7) + (1 \times 6) + (7 \times 5) + (4 \times 4) + (9 \times 3) + (2 \times 2) = 200$

$200 + 9 = 209$

(b) $(1 \times 10) + (8 \times 9) + (6 \times 8) + (2 \times 7) + (8 \times 6) + (2 \times 5) + (0 \times 4) + (6 \times 3) + (2 \times 2) = 224$

$224 + 7 = 231$

(c) $(0 \times 10) + (8 \times 9) + (5 \times 8) + (3 \times 7) + (2 \times 6) + (1 \times 5) + (1 \times 4) + (4 \times 3) + (7 \times 2) = 180$

$180 + 7 = 187$

(d) $(0 \times 10) + (8 \times 9) + (6 \times 8) + (3 \times 7) + (1 \times 6) + (3 \times 5) + (5 \times 4) + (8 \times 3) + (4 \times 2) = 214$

$214 + 6 = 220$

Barcodes -1 .... page 54

1. (a) 9 ③①⓪⓪⑤⑤ ①④⑨⑧⓪ ⑨

$3 + 0 + 5 + 1 + 9 + 0 = 18$

$18 \times 3 = 54$

$9 + 1 + 0 + 5 + 4 + 8 = 27$

$54 + 27 = 81$

$81 + 9 = 90$

When the check digit in the square is added, the total is 90, i.e. 81 + 9 = 90. 90 is a multiple of ten, so the barcode is correct.

(b) 9 ③⓪⓪⑥⑤⑦ ⓪①⑤③③ ⑥

$3 + 0 + 5 + 0 + 5 + 3 = 16$

3 x 16 = 48

9 + 0 + 6 + 7 + 1 + 3 = 26

48 + 26 = 74

74 + 6 = 80

80 is a multiple of ten so the barcode is correct.

2. (a) 9 300657 30446 1

3 + 0 + 5 + 3 + 4 + 6 = 21

3 x 21 = 63

9 + 0 + 6 + 7 + 0 + 4 = 26

63 + 26 = 89

89 + 1 = 90

90 is a multiple of ten so the barcode is correct.

(b) 9 300652 01075 6

3 + 0 + 5 + 0 + 0 + 5 = 13

3 x 13 = 39

9 + 0 + 6 + 2 + 1 + 7 = 25

39 + 25 = 64

64 + 6 = 70

70 is a multiple of ten so the barcode is correct.

(c) 9 310060 00623 4

3 + 0 + 6 + 0 + 6 + 3 = 18

3 x 18 = 54

9 + 1 + 0 + 0 + 0 + 2 = 12

54 + 12 = 66

66 + 4 = 70

70 is a multiple of ten so the barcode is correct.

Barcodes - 2 .... page 55

1. Teacher check

2. (a) Food products have barcodes that are scanned to show their price.

(b) The checkout person will type the 13 digits into the register to reveal the price.

(c) The barcodes are marked with a felt-tipped marker so that the person buying them is not charged the full price. The checkout person will stop and look at the product to find the new discounted price.

Pascal's Triangle - 1 .... page 56

1.

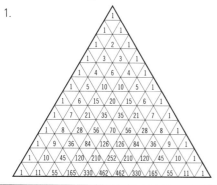

Note: Each new row is formed by each pair of numbers from the previous row. Each row begins and ends with one.

2. Various patterns may be observed along the diagonals.

Diagonal 1: All the number 1s.

Diagonal 2: 1, 2, 3, 4, 5, 6, 7, 8

Diagonal 3: 1, 3, 6, 10, 15, 21, 28 (triangular numbers).

3. (a) 121   1,331

(b) The answers are the same as the numbers in the third and fourth rows of Pascal's Triangle.

(c) Teacher check 11^4 = 14,641

(d) The number 161,051 is formed which is not the same as the next row but does bear some relationship to it if you look closely.

Pascal's Triangle - 2 .... page 57

1.

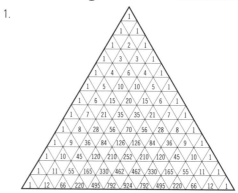

2. 1, 2, 4, 8, 16   3. 32, 64, 128

4. Each number is double the previous number.

5. (a) 220   (b) The totals are the same.

2^0	2^1	2^2	2^3	2^4	2^5	2^6	2^7	2^8	2^9	2^{10}
1	2	4	8	16	32	64	128	256	512	1,024

6. Adding the total for each row on Pascal's Triangle gives the answers: 1 is 2^0, row 2 is 2^1, row 3 is 2^2, row 4 is 2^3, row 5 is 2^4 and so on.

Coloring Pascal's Triangle - 1 .... page 58

1.

2.

3. When the multiples of five are colored, three upside down triangles are formed and the beginnings of three more start to show. The pattern is symmetrical.

Coloring Pascal's Triangle - 2 .... page 59

1.

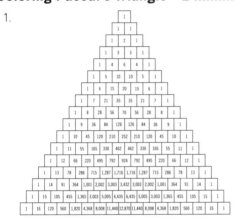

2.

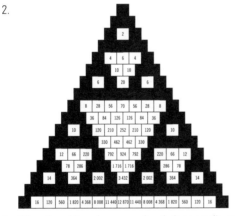

3. When the odd numbers are colored, three small upside down triangles surround one large upside down triangle.

Fun with Fibonacci .... page 60

1. Each number in the sequence is formed by adding the previous two numbers in the sequence.

2. 55, 89, 144

3. (a) 1   (f) 1.625

   (b) 2   (g) 1.6153846

   (c) 1.5   (h) 1.6190476

(d) 1.6666667 (really 1.6)

(i) Teacher check

(e) 1.6   (j) Teacher check

4. The numbers get close to and hover around 1.6.

5. The result always differs by one.

challenge A typical small size envelope conforms to the Golden Ratio.

Fibonacci Facts .... page 61

1. 1, 1, 2, 3, 5, 8, 13, 21, 34, 55, 89, 144

2. (a) 54. One less than the tenth Fibonacci number, 55.

   Adding the first nine Fibonacci numbers produces a total (88) – one less than the eleventh Fibonacci number.

   (b) Adding the first ten Fibonacci numbers produces a total one less than the twelfth Fibonacci number and so on.

3. 1 + 2 + 5 = 8, 1 + 2 + 5 + 13 = 21,

   1 + 2 + 5 + 13 + 34 = 55

   Adding every second term produces the next Fibonacci number in the sequence.

4. 1 + 3 + 8 = 12, 1 + 3 + 8 + 21 = 33,

   1 + 3 + 8 + 21 + 55 = 88

   Using this sequence the result is always one less than the next Fibonacci number.

5. 1, 3, 4, 7, 11, **18, 29, 47, 76, 123, 199**

Golden Rectangles .... page 62

1. Teacher check

2. Many objects measured will produce the Golden Ratio.

   Note: The Golden Ratio is referred to by the symbol Phi.

The Golden Ratio .... page 63

1. Length is 160 mm, width 100 mm

   160 ÷ 100 = 1.6

   1.6 is close to the Golden Ratio.

2.-3. Each successive rectangle that is formed will conform to the Golden Ratio.

Prime Time .... page 64

1.

2. The prime numbers seem to run in diagonals from right to left.

3. The diagonal in which the prime numbers lie are either side of the multiples of six.

4. (a) 3 + 5 = 8, 3 + 7 = 10, 5 + 7 = 12, 3 + 11 = 14, 5 + 11 = 16,

Patterns in Mathematics

7 + 11 = 18

(b) 7 + 13 = 20, 3 + 17 = 20, two ways.

(c) 5 + 37 = 42, 11 + 31 = 42, 13 + 29 = 42, 19 + 23 = 42, four ways.

Number Patterns - 1 ....page 65

1. (a) 20, 24, 28 – adding four each time.

   (b) 21, 25, 29 – adding four each time.

   (c) 75, 71, 67 – subtracting four each time.

   (d) The number sequence increases or decreases by four each time.

   (e) The first two sequences increase by four and the final sequence decreases by four.

2. (a) 81, 234, 729 – multiply by three.

   (b) 48, 96, 192 – multiply by two.

   (c) 10,000, 100,000, 1,000,000 – multiply by ten.

   (d) The numbers increase by a multiplication of one number each time.

   (e) The three sequences are multiplied by different numbers, e.g. 3, 2, 10.

3. (a) 46, 94, 190 – multiply the previous number by two and add two.

   (b) 31, 63, 127 – multiply the previous number by two and add one.

   (c) 17, 33, 65 – multiply by two and subtract by seven.

   (d) The number sequences are formed by multiplying by two and then adding or subtracting a number.

   (e) The second operation (e.g. +2, +1, −1) is different.

4. Teacher check

Number Patterns - 2 ....page 66

1. 21, 25, 29 – adding 4 each time.

2. 49, 43, 37 – subtracting 6 each time.

3. 22, 29, 37 – adding 3, then 4, then 5; i.e. one more each time a number is added.

4. 26, 37, 50 – starting with adding 3, then 5, then 7; i.e. adding two more than the previous number added.

5. 32, 64, 128 – doubling pattern.

   Powers of two – 2^0, 2^1, 2^2, 2^3, 2^4, 2^5, 2^6 …

6. 75, 90, 105 – adding 15 each time.

7. 12, 6, 3 – halving or dividing by two.

8. 81, 243, 729 – multiplying by three

9. 36, 49, 64 – square numbers.

10. 0.9, 1.0, 1.1 – adding 0.1 each time.

11. 0.1, 0.01, 0.001 – dividing by ten or multiplying by one-tenth.

12. 13, 21, 34 – Fibonacci sequence.

    Add the two previous terms in the sequence.

13.-14. The purpose of this question is to illustrate that three terms is not enough to establish a sequence. Several possibilities exist,

    e.g. 1, 2, 4, 7, 11, 16 … (i.e. adding one, then two, then three and so on.)

    1, 2, 4, 8, 16, 32 – doubling each time or powers of two.

Super Sequences ....page 67

1. (a) 1, 2, 3, 4, 1, 2, 3

   (b) Four

   (c) The pattern is running in cycles of four. i.e. the fourth, eighth, twelfth digits are always four, therefore, the twentieth digit will be four.

   (d) Three

   (e) Using similar reasoning, the fortieth number in the sequence will be four, so the thirty-ninth must be three.

   (f) 11

   (g) There is one 2 in every group of four, so in every forty numbers there will be ten 2s. Two will occur once more in the remaining three numbers.

2. (a) 20

   (b) Single digit – one seven required; Two-digit numbers – one per decade except 70s – 8 required;

   Seventies – 11 required because 77 required two 7s.

   (c) Twenty-one of number 1; Twenty of all the other digits except zero; Eleven zeros.

Power Patterns ....page 68

1. (a) 1, 6, 36, 216, 1,296, 7,776, 46,656, 279,936, 1,679,616, 10,077,696

   (b) The last digit is always 6.

   (c) 6 6^{10} = 60,446,176

   (d) Teacher check

2. (a)

2^x	Result	Last Digit	3^x	Result	Last Digit	4^x	Result	Last Digit
2^1	2	2	3^1	3	3	4^1	4	4
2^2	4	4	3^2	9	9	4^2	16	6
2^3	8	8	3^3	27	7	4^3	64	4
2^4	16	6	3^4	81	1	4^4	256	6
2^5	32	2	3^5	243	3	4^5	1,024	4
2^6	64	4	3^6	729	9	4^6	4,096	6
2^7	128	8	3^7	2,187	7	4^7	16,384	4
2^8	256	6	3^8	6,561	1	4^8	65,536	6
2^9	512	2	3^9	19,683	3	4^9	262,144	4
2^{10}	1,024	4	3^{10}	59,049	9	4^{10}	1,048,576	6

Note: 2^{10} and 4^5 produce the same result

Answers

A Million Dollars Per Month .... page 69

1. Teacher check

   Note: While a $1,000,000 a month sounds like a large amount, it pales into insignificance compared to the second option.

2. The following table shows how quickly the payments increase under option 2.

Day	Payment in cents	Total payment in cents
1	$1 (2^0)$	$(2^1 - 1) = 1$
2	$2 (2^1)$	$(2^2 - 1) = 3$
3	$4 (2^2)$	$(2^3 - 1) = 7$
4	$8 (2^3)$	$(2^4 - 1) = 15$
5	$16 (2^4)$	$(2^5 - 1) = 31$
6	$32 (2^5)$	$(2^6 - 1) = 63$
7	$64 (2^6)$	$(2^7 - 1) = 127$
8	$128 (2^7)$	$(2^8 - 1) = 255$
9	$256 (2^8)$	$(2^9 - 1) = 511$
10	$512 (2^9)$	$(2^{10} - 1) = 1,024$
11	$1,024 (2^{10})$	$(2^{11} - 1) = 2,047$
12	$2,048 (2^{11})$	$(2^{12} - 1) = 4,095$
13	$4,096 (2^{12})$	$(2^{13} - 1) = 8,192$
14	$8,192 (2^{13})$	$(2^{14} - 1) = 16,383$
15	$16,384 (2^{14})$	$(2^{15} - 1) = 32,767$
16	$32,768 (2^{15})$	$(2^{16} - 1) = 65,535$
17	$65,536 (2^{16})$	$(2^{17} - 1) = 131,071$
18	$131,072 (2^{17})$	$(2^{18} - 1) = 262,143$
19	$262,144 (2^{18})$	$(2^{19} - 1) = 524,287$
20	$524,288 (2^{19})$	$(2^{20} - 1) = 1,048,575$
21	$1,048,576 (2^{20})$	$(2^{21} - 1) = 2,097,151$
22	$2,097,152 (2^{21})$	$(2^{22} - 1) = 4,194,303$
23	$4,194,304 (2^{22})$	$(2^{23} - 1) = 8,388,607$
24	$8,388,608 (2^{23})$	$(2^{24} - 1) = 16,777,215$
25	$16,777,216 (2^{24})$	$(2^{25} - 1) = 33,554,431$
26	$33,554,432 (2^{25})$	$(2^{26} - 1) = 67,108,863$
27	$67,108,864 (2^{26})$	$(2^{27} - 1) = 134,217,727$
28	$134,217,728 (2^{27})$	$(2^{28} - 1) = 268,435,455$
29	$268,435,456 (2^{28})$	$(2^{29} - 1) = 536,870,911$
30	$536,870,912 (2^{29})$	$(2^{30} - 1) = 1,073,741,823$

   After 27 days you will have been paid over $1.3 million dollars. On day 28 you will receive over another $1.3 million dollars pay.

3. In a 30-day month, earnings would total 2^{30} cents or 1,073,741,823¢ or $10,737,418.23. The second option pays about 10 times the first.

Ordered Pairs .... page 70

1. (a) (17, 15), (16, 14), ... (10, 8)

   Subtract two from the first number.

   (b) (4, 12), (5, 15), ... (22, 66)

   The second number is three times the first.

   (c) (37, 27), (36, 26), ... (10, 0)

   The second number is ten less than the first.

   (d) (6, 36), (5, 25), ... (20, 400)

   A squaring pattern relates the pairs.

   (e) (4, 36), (5, 45), ... (9, 81)

   A factor of nine relates each number in the pair.

   (f) (10, 26), (11, 27), ... (75, 91)

   Adding 16 to the first number in the pair produces the second number.

   (g) (39, 36), (38, 35), ... (3, 0)

   Subtracting three from the first number produces the second number in the pair.

2. (a)

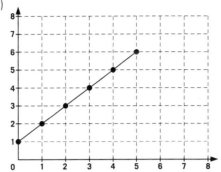

   (b) When the ordered pairs are graphed a line is formed.

   (c) (6, 7), (7, 8)

Paths and Patterns .... page 71

1.

Shape	Regions (R)	Vertices (V)	Paths or Edges (E)	R + V − E
(a)	3	4	5	2
(b)	3	2	3	2
(c)	2	6	6	2
(d)	3	8	9	2
(e)	3	6	7	2
(f)	1	4	3	2
(g)	5	5	8	2
(h)	2	3	3	2

2. Clearly, a relationship exists between the regions, vertices and edges of a network.

3. $R + V - E = 2$ or $R + V - 2 = E$

Circle Patterns .... page 72

1.    2.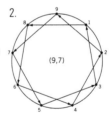

   (9,2)   (9,7)

3. The shapes are the same.

4. (a) 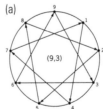   (b)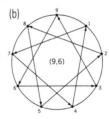

   (9,3)   (9,6)

5. (a) Teacher check

   (b)

   (9,4)   (9,5)

Answers

1.

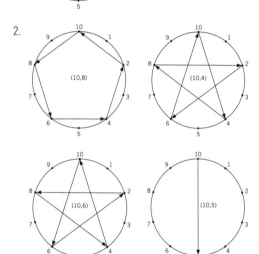

2.

3. Teacher check

4. Teacher check